Lecture Notes in Mathematics

A collection of informal reports and seminars
Edited by A. Dold, Heidelberg and B. Eckmann, Zürich

103

M. F. Atiyah, J. Eells, K. M. Hoffman,
L. Hörmander, C. E. Rickart, H. Rossi,
S. Smale, F. Treves, J. Wermer

Lectures in Modern Analysis and Applications I

Edited by C. T. Taam, George Washington University

Springer-Verlag Berlin Heidelberg GmbH

DOI 10.1007/978-3-540-36145-9

MyCopy version of the original edition 1969

www.springer.com/mycopy

PREFACE

This lecture series in Modern Analysis and Applications was sponsored by the Consortium of Universities (American University, Catholic University, Georgetown University, George Washington University, and Howard University) in Washington, D. C. and the University of Maryland, in conjunction with the U. S. Air Force Office of Scientific Research. These lectures were presented at the sponsoring universities over the period 1967-69 by mathematicians who have contributed much to the recent growth of analysis.

The series consisted of eight (8) sessions of three lectures each. Each session was devoted to an active basic area of contemporary analysis which is important in application or shows potential applications. Each lecture presented a survey and a critical review of certain aspects of that area, with emphasis on new result, open problem, and application.

The eight sessions of the series were devoted to the following basic areas of analysis:

1. Modern Methods and New Results in Complex Analysis
2. Banach Algebras and Applications
3. Topological Linear Spaces and Applications
4. Geometric and Qualitative Aspects of Analysis
5. Analysis and Representation Theory
6. Modern Analysis and New Physical Theories
7. Modern Harmonic Analysis and Applications
8. Integration in Function Spaces and Applications.

This volume contains nine lectures from the first four sessions. Other lectures will appear in subsequent volumes.

C. T. Taam

George Washington University

ORGANIZING COMMITTEE

George H. Butcher
Howard University

Avron Douglis
University of Maryland

John E. Lagnese
Georgetown University

Raymond W. Moller
Catholic University of America

Robert G. Pohrer
U. S. Air Force Office of Scientific Research

Steven H. Schot
American University

C. T. Taam, Chairman
George Washington University

Elmer West
Consortium of Universities

TABLE OF CONTENTS

MODERN METHODS AND NEW RESULTS IN COMPLEX ANALYSIS

Professor CHARLES E. RICKART, Yale University

A function algebra, which does not contain all continuous functions, may exhibit certain properties reminiscent of analyticity. An example is the local maximum modulus principle proved by Hugo Rossi. This, along with various other results, suggests the beginnings of an abstract analytic function theory. At this stage, the program is to obtain analogues of certain results from several complex variables.

TOPOLOGICAL LINEAR SPACES AND APPLICATIONS

Professor LARS HÖRMANDER, Institute for Advanced Study

For the Cauchy problem with data on a hyperplane there exists a unique solution for arbitrary data if and only if the equation is hyperbolic in the sense of Garding. When the hyperplane is characteristic there is no longer uniqueness, but we characterize the equations having a solution for arbitrary Cauchy data. This class contains all parabolic equations.

Professor F. TREVES, Purdue University

Local Cauchy problems for systems of linear PDEs with analytic coefficients, with data on noncharacteristic hypersurfaces, have always unique solutions. But these in general need not be distributions, they are ultradistributions. A simple proof of this fact is possible, based on general results about abstract differential equations (also valid for nonlinear ones), suitably adapted blowing up of small domains and ladders of functional (Banach) spaces. This allows a detailed description of the situation, including of the symbols of "fundamental solutions", and reveals the links with the problem of solvability in more classical sense.

GEOMETRIC AND QUALITATIVE ASPECTS OF ANALYSIS

* This lecture was presented by Professor James Eells.

Bounded Analytic Functions in the Unit Disk

by

Kenneth Hoffman

This talk will be entirely expository. We shall be concerned with the boundary behavior of some analytic functions. The results to be discussed can be found principally in two papers, one by Carleson [1], the other by the speaker [4].

Let

$$D = \{z \in C ; |z| < 1\}$$

be the open unit disc in the plane. Let H^∞ (the limiting Hardy space) be the family of bounded analytic functions on D . At points interior to D , the functions in H^∞ are (of course) very well-behaved. Near the boundary of D , their behavior can be somewhat wild; and there are many delicate and substantial problems about boundary behavior. The study of those problems has a long history, which we cannot go into; however, we shall mention one or two things.

In 1906, Fatou [3] proved the following positive result about boundary values. If $f \in H^\infty$, then the radial limit

$$f(e^{i\theta}) = \lim_{r \to 1} f(re^{i\theta})$$

exists for almost every θ , and we may recapture f from the boundary values by the Cauchy formula

$$f(z) = \frac{1}{2\pi i} \int_{|\lambda|=1} \frac{f(\lambda)}{\lambda - z} d\lambda .$$

In spite of the Fatou theorem, bounded analytic functions may behave erratically near the boundary. We can illustrate that by considering a class of functions known as Blaschke products.

Let $\{\alpha_n\}$ be any sequence in D such that

$$\sum_n (1 - |\alpha_n|) < \infty .$$

The infinite (Blaschke) product

$$B(z) = \prod_n \frac{\bar{\alpha}_n}{|\alpha_n|} \frac{\alpha_n - z}{1 - \bar{\alpha}_n z}$$

converges, uniformly on compact subsets of D, and defines a function B in H^∞. Evidently $|B| < 1$ and $\{\alpha_n\}$ is the sequence of zeros of B. It is not difficult to show that (for a Blaschke product) the radial limits

$$B(e^{i\theta}) = \lim_{r \to 1} B(re^{i\theta})$$

have absolute value 1 at almost every point of the unit circle. The sequence $\{\alpha_n\}$ could be dense on the unit circle. In such a case, $B(z)$ must oscillate rather wildly as z approaches the boundary.

Classically, much of the work on boundary behavior has

concentrated on cluster sets. There are several elegant results about cluster sets of bounded analytic functions. For those we refer to the recent book of Collingwood and Lohwater [2]. The point of view which we adopt here is somewhat different. We compactify the disc D in such a way that each bounded analytic function admits a continuous extension to the compactification. Then, we can study not only the limit values at the boundary but also the geometry of the compact space on which those values are assumed.

There is a simple and direct way to find the appropriate compactification in this case, because H^∞ is an algebra, i.e., sums and products of bounded analytic functions are bounded analytic functions. We follow the method which Gelfand employed for Banach algebras. For details, see [5; Chapter 10]. We let

$$M = \hom(H^\infty;C)$$

be the set of all homomorphisms from the algebra H^∞ onto the algebra of complex numbers. With each $f \in H^\infty$ we associate a complex-valued functions $\hat{f}$ on M by

$$\hat{f}(m) = m(f) .$$

Each point $\alpha \in D$ defines a complex homomorphism m_α on H^∞ by

$$m_\alpha(f) = f(\alpha)$$

so that

$$\hat{f}(m_\alpha) = f(\alpha) .$$

Thus, if we agree to identify α with m_α, we may regard the disc D as a subset of M; and, for each f in H^∞, $\hat{f}$ is an extension to M of the function f on the subset D.

Now, we endow M with a topology--the weakest (smallest) topology on M which makes every function $\hat{f}$ continuous. With that topology, M is a compact Hausdorff space and D is an open subset of M. Then we have essentially what we were after, because, virtually by definition, each bounded analytic function on D has a continuous extension to M. A priori, this space M is the largest compact Hausdorff space on which H^∞ can be realized as a point-separating algebra of continuous functions. The question naturally arises Is M a compactification of D, i.e., is D dense in M? About 1957, considerable interest in that problem was aroused, and the conjecture that D is dense in M became known as the "corona conjecture".

The corona conjecture sounds somewhat abstract; however, it can be translated rather easily into a "classical" conjecture about bounded analytic functions: If $f_1, \cdots, f_n$ are functions in H^∞ such that $|f_1| + \cdots + |f_n| \geq \delta > 0$, there exist bounded analytic functions $g_1, \cdots, g_n$ such that

$$f_1 g_1 + \cdots + f_n g_n = 1 .$$

The conjecture was finally proved in this classical form by Carleson [1]. We should also acknowledge (as does Carleson) the significant contribution made to the proof by D. J. Newman. We shall not attempt to go into Carleson's proof. It is a difficult and technical argument. But we should mention that the crux of the matter is the analysis of sets $\{z \; ; \; |B(z)| < \varepsilon\}$ for certain Blaschke products B . The corona theorem must be regarded as one of the deepest results known concerning bounded analytic functions.

We shall now discuss another question concerned with the behavior of the extended functions $\hat{f}$ on the "fringe" $M - D$. We ask ourselves to find all analytic subsets of M . An <u>analytic set</u> S in M is a set which is the image of a map

$$V \xrightarrow{\tau} S$$

where

(a) V is a connected analytic space;

(b) τ is an analytic map, in the sense that the composition $\hat{f} \circ \tau$ is analytic on V , for every f in H^∞ .

Briefly, an analytic set is a subset which can be endowed with an analytic structure, relative to which all of the functions $\hat{f}$ are analytic. The precise definition of analytic space or structure is not very important, because it will follow from the results we are about to discuss that all analytic sets in M are contained in analytic discs.

The question about analytic sets is not an idle one. It cannot be translated directly into a classical question; however, it is a basic and natural question, from a certain viewpoint. Here, there is time only to comment briefly on that. Suppose A is an algebra of continuous complex functions on a compact Hausdorff space M , and suppose it is <u>not</u> possible to approximate each continuous function uniformly by functions in A . Every known case of such an algebra can be explained by the fact that the functions in A are analytic in one or another sense. There is a vague, but basic question which asks if analyticity explains every such algebra A . As part of an assault on that question, it is of interest to classify all analytic sets in M , for the specific algebras A which we know.

For the algebra H^{∞} , one analytic subset of M is the disc D . Thus the interest is in finding those analytic sets which lie in the fringe M - D . On the surface, it is not at all clear that there are any non-trivial sets in M - D Probably we would not be talking about analytic sets, if there were none, other than points. I. J. Schark [6] showed that analytic discs exist in $M - D$. That is, he showed that there exists a map

$$\mathcal{D} \xrightarrow{\tau} M - D$$

where

(a) $\mathcal{D}$ is an open disc in the plane;

(b) τ is 1:1 and analytic.

In fact, there exist a great many such analytic discs in $M - D$, and one can describe them in various ways. There also exist many points in $M - D$ which do not lie on any analytic disc.

The principal result of the speaker's paper [4] was the proof of the following. Let m be a point of the space M . These are equivalent.

(1) There exists in M a non-trivial analytic set which contains m .

(2) The ideal of functions which vanish at m is different from its own square; that is, not every $f \in H^\infty$ which vanishes at m is the product of two such functions.

(3) If S and T are subsets of the disc D each of which captures m in its closure, then the hyperbolic distance from S to T is zero.

(4) m lies in the closure of a sequence $\{\alpha_n\}$ in D which satisfies the condition

$$0 < \inf_n \prod_{k \neq n} \left| \frac{\alpha_k - \alpha_n}{1 - \bar{\alpha}_k \alpha_n} \right| .$$

The equivalence of (1)-(4) is established by means of a factorization theorem for Blaschke products. It is interesting to note that the proof of the equivalence is concerned (as was the proof of the corona conjecture) with the set $\{z ; |B(z)| < \varepsilon\}$, where B is a Blaschke product.

It is also possible to show that, if there is any non-trivial analytic set through the point m, then there is a largest such set and it has the structure of an analytic disc. One can also describe how to obtain the maps which coordinatize those analytic discs. We shall close by describing the process roughly, because it helps to render the discs in $M - D$ somewhat more tangible.

Suppose that m is a point of M which lies in the closure of a sequence $\{\alpha_n\}$ as in (4) about

$$\sum_n (1 - |\alpha_n|) < \infty ,$$

and so we have the associated Blaschke product B. The condition in (4) states precisely that

$$0 < \inf_n (1 - |\alpha_n|^2) \, |B'(\alpha_n)| .$$

That derivative condition guarantees that, for small positive ε, the set $\{z ; |B(z)| < \varepsilon\}$ will break up into a sequence of simply connected regions R_n. Each R_n will contain just one zero α_n of B . In fact B maps R_n conformally onto the disc of radius ε about the origin. Out of $\{\alpha_n\}$ we can extract some sort of generalized sequence which converges in M to the point m . The idea is that, as that convergence takes place, the corresponding regions (about the α's) converge to a "region" through the point m .

Time has run out on us. Hopefully, these brief remarks have given some indication of the blending of "modern" methods and concepts with the classical theory of analytic functions.

Bibliography

1. Carleson, L., Interpolation by bounded analytic functions and the corona problem, Ann. of Math., 76 (1962), 542-559.

2. Collingwood, E. and Lohwater, A. J., The Theory of Cluster Sets, Cambridge University Press, Cambridge, 1967.

3. Fatou, P., Series trigonometriques et series de Taylor, Acta Math., 30 (1906), 335-400.

4. Hoffman, K., Bounded analytic functions and Gleason parts, Ann. of Math., 86 (1967), 74-111.

5. __________, Banach Spaces of Analytic Functions, Prentice-Hall, Englewood Cliffs, 1962.

6. Schark, I. J., Maximal ideals in an algebra of bounded analytic functions, J. Math. Mech., 10 (1961), 735-746.

Strongly Pseudoconvex Manifolds

by

Hugo Rossi[1]

1. Levi's problem.

Let D be a domain (open connected set) in $\mathbb{C}^n$. Is it possible to find a function f, holomorphic in D so that the boundary of D is the natural boundary of f? If $n=1$ this is easy to do, but for $n>1$ it has long been known that there are domains that do not admit such a function. Those domains which are the "natural domain of definition" of some holomorphic function are called domains of holomorphy. It became an important problem to give geometric or other function-theoretic characterizations of such domains. In 1951, in the Seminaire de l'Ecole Normale Superieure [5], as the culmination of half a century's investigations (by Behnke, H. Cartan, Oka and many others), H. Cartan gave ten such characterizations. The study of strongly pseudoconvex manifolds is an outgrowth of one of those properties.

This study began early in the twentieth century with the work of E. E. Levi [24]. He sought a characterization of domains of holomorphy in terms of the boundary alone.

1 This work was partially supported by Army Grant No. DA-ARO-D-31-124-G866.

He considered domains $D \subset \mathbb{C}^2$, which have a differentiable boundary. We can write such domains in the form $D = \{x \in \mathbb{C}^2; \varphi(x) < 0\}$ where φ is a differentiable function in $\mathbb{C}^2$, and $d\varphi(x) \neq 0$ whenever $\varphi(x) = 0$. E. E. Levi observed that if D is a domain of holomorphy, then for any $x_0 \in bD$, the boundary of D, this implication holds:

$$(1) \qquad \frac{\partial\varphi}{\partial z_1}(x_0)\cdot t^1 + \frac{\partial\varphi}{\partial z_2}(x_0)\, t^2 = 0 \quad \text{implies} \quad \Sigma \frac{\partial^2\varphi}{\partial z_i \partial \bar{z}_j}(x_0)\, t^i \bar{t}^j \geq 0.$$

Conversely, if x_0 is any boundary point of a domain D such that (1) holds with the strict inequality for $(t^1, t^2) \neq (0,0)$, then there is a ball B with center x_0 such that $B \cap D$ is a domain of holomorphy. Any domain such that (1) holds for all x_0 on the boundary is called pseudoconvex. It became known as Levi's problem to prove that a pseudoconvex domain is a domain of holomorphy. This was accomplished in 1942 by Oka [31] for domains in $\mathbb{C}^2$. The generalization to n variables was performed in 1951 by Bremermann, Norguet and Oka [3,30,31] independently. In 1958 Hans Grauert reconsidered Levi's problem in a more general setting [8], as a step in his proof of the imbedding of real-analytic manifolds. The present report properly begins with this work of Grauert.

2. Grauert's solution of Levi's problem.

In the intervening years the problem had undergone considerable generalization. As manifolds entered the mathematical picture, K. Stein generalized the notion of a domain of holomorphy.

Definition 1. A complex manifold X is called *holomorphically convex* if for every closed discrete sequence $\{x_n\} \subset X$, there exists a holomorphic function f such that the set $\{f(x_n)\}$ is unbounded. A holomorphically convex manifold is called a *Stein manifold* if in addition these two conditions are satisfied:

(i) for all $x \neq y$ in X there is an f holomorphic on X such that $f(x) \neq f(y)$.

(ii) for all $x \in X$, there is a holomorphic map $F: X \longrightarrow \mathbb{C}^n$ such that $dF(x)$ is nonsingular.

Levi's notion of pseudoconvexity generalizes as follows.

Definition 2. Let X be a complex manifold. $M \subset X$ is called a *strongly pseudoconvex manifold* if M is a relatively compact domain in X and we can write $M = \{x \in X;\ \varphi(x) < 0\}$, where φ is a differentiable function with this behavior on bM:

(i) for $x_o \in bM$, $d\varphi(x_o) \neq 0$,

(ii) If $z^1,\dots,z^n$ are local coordinates at $x_o \in bM$, the hermitian quadratic form $\left(\frac{\partial^2 \varphi}{\partial z_i \partial \bar{z}_j}(x_o) \right)$ is positive definite.

Grauert proved that a pseudoconvex manifold is holomorphically convex. He would not have expected it to be Stein because of the many counterexamples, arising principally from algebraic geometry. Consider the following. Let $\mathbb{C}^* = \mathbb{C} - \{o\}$. Let two copies of $\mathbb{C} \times \mathbb{C}$ be given with coordinates (z,ξ), (w,η) respectively. Paste these two copies together along $\mathbb{C}^* \times \mathbb{C}$ so that (z,ξ) corresponds to (w,η) when $w = 1/z$ and $\eta = z\xi$. The resulting manifold X is covered by two coordinate neighborhoods U_1, U_2 with coordinates (z,ξ), (w,η) respectively, and the transition map is

$$w = 1/z$$

$$\eta = z\xi$$

Let $M = \{(z,\xi) \in U_1;\ |z\xi|^2 + |\xi|^2 < 1\} \cup \{(w,\eta) \in U_2;\ |w\eta|^2 + |\eta|^2 < 1\}$.

On the overlap the two inequalities are the same, so M is a manifold of the type $\{x \in X;\ \varphi(x) < 0\}$ with φ differentiable. It is easily checked that φ has the required properties, so that M is strongly pseudoconvex. However the set $\{(z,\xi)\in U_1;\ \xi=0\}\cup\{(w,\eta)\in U_2;\ \eta=0\}$ is a compact positive-dimensional submanifold of M, so by the maximum principle condition (i) for M to be Stein cannot be satisfied.

We will see below that this is typical of a strongly pseudoconvex manifold: if it is not Stein then it contains at most finitely many positive dimensional compact subvarieties.

Grauert's solution of Levi's problem combine the techniques of sheaf theory with the theory of compact operators on Frechet spaces (the relevant compactness is that of uniformly bounded families of holomorphic functions). Sheaf theory has been developed by H. Cartan and J. Leray in the early 1950's [5] out of the work of Oka and it had become apparent that many of the important problems of complex analysis were best formulated in the cohomology theory of coherent analytic sheaves. The following basic theorem of H. Cartan [5] was the starting point.

Theorem 1. (Cartan's theorem B). Let X be a complex manifold. X is Stein if and only if, for every coherent analytic sheaf S on X, $H^p(X,S)=0$ for $p>0$.

(The difficult part of the theorem is of course the only if statement).

The crucial part of Grauert's work was the following generalization of Cartan's theorem B [8].

Theorem 2. (Grauert). Let M be a strongly pseudoconvex submanifold of the complex manifold X. Let S be a coherent analytic sheaf on X. Then $H^p(M,S)$ is a finite dimensional complex vector space, for $p > 0$.

This is indeed a generalization of theorem B for that theorem can be easily deduced from this. In fact theorem 2 can be proven directly, without appeal to theorem B, thus giving an alternative proof of Cartan's theorem (cf., for example [33]).

We shall here use theorem 2 (and the holomorphic convexity of M) to give a geometric description of strongly pseudoconvex manifolds. (Some of these results are very much like some work of Narasimhan [27], but the techniques involved are different). These results are not entirely new, but represent a simplification of previous proofs.

3. The Support of cohomology.

Since the results of this section do not appear elsewhere, we shall indicate proofs where they are new.

For M a complex manifold, let $\mathcal{O}(M)$ be the ring of functions holomorphic on X. By a level set we mean a set of the form

$$L_{x_0} = \{x \in M; \ f(x) = f(x_0) \ , \ \text{for all } f \in \mathcal{O}(M)\} \ ,$$

for some point $x_0 \in X$. Clearly level sets are subvarieties of M. If M is strongly pseudoconvex, these level sets are compact. For if L_{x_0} is not compact, it contains a closed discrete sequence. But by the holomorphic convexity there is an $f \varepsilon \mathcal{O}(M)$ which is unbounded, let alone constant, on this sequence. Now if L is any compact subvariety of M, let L' be the union of the positive dimensional branches of L. We want to prove that there are at most finitely many such L' in all.

This can be proven directly from the holomorphic convexity via the proper mapping theorem (that is to say, it is a geometric, rather than a sheaf-theoretic fact). Nevertheless the phenomenon of non-vanishing cohomology remains unexplained except that is caused by the presence of these compact varieties. Here we shall directly attack the cohomology of a coherent sheaf and explicitly relate it to these varieties. The following tool is essential:

Lemma. Let $M = \{x \varepsilon X;\ \varphi(x) < 0\}$ be strongly pseudoconvex in X. There is an $\varepsilon_0 > 0$ such that if K is a positive dimensional compact subvariety of $\{x \varepsilon X;\ \varphi(x) < \varepsilon_0\}$, then $K \subset \{x \varepsilon X;\ \varphi(x) < -\varepsilon_0\}$.

Proof. This lemma says that the positive dimensional parts of the level sets do not come too near to the boundary of M. Now φ is at least twice continuously differentiable and is strongly plurisubharmonic along bM. Thus there is an $\varepsilon_0 > 0$ such that if $|\varphi(x)| \leq \varepsilon_0$, φ is strongly plurisubharmonic at x. Now if K is a connected, compact, positive dimensional subvariety of X, φ assumes a maximum on K. This maximum point cannot be a point of strong plurisubharmonicity of φ because of the maximum principle (cf p.272 of [12]). Thus the lemma is verified.

Let $M' = \{x \in X;\ \varphi(x) < \varepsilon\}$ with $|\varepsilon| < \varepsilon_0$. M' is also strongly pseudoconvex, and by this lemma the positive dimensional parts of the level sets of $\mathcal{O}(M')$ are finite unions of the compact positive dimensional subvarieties of M. With this observation made, we may proceed to the main result.

Theorem 3. Let M be a strongly pseudoconvex manifold in X. Let S be a coherent analytic sheaf on X and $p > 0$. There are functions $f_1, \ldots, f_t \in \mathcal{O}(M)$ such that

(i) $V = \{x \in M;\ f_i(x) = 0,\ 1 \leq i \leq t\}$ consists of finitely many level sets,

(ii) for I the sheaf of ideals generated by $f_1, \ldots, f_t$ the restriction map $H^p(M,S) \longrightarrow H^p(M, \mathcal{O}/I \otimes_{\mathcal{O}} S)$ is injective.

Remark. $H^p(M, \mathcal{O}/I \otimes S) = H^p(V', \mathcal{O}/I \otimes_{\mathcal{O}} S)$ has support on a finite union of positive dimensional compact subvarieties.

Proof. Let $M = \{x \in X;\ \varphi(x) < 0\}$, and $M' = \{x \in X;\ \varphi(x) < \varepsilon\}$, where $0 < \varepsilon < \varepsilon_0$, and ε_0 is given by the lemma. Both M,M' are strongly pseudoconvex and $\overline{M}$ is compact in M'.

Now if S is a coherent sheaf of $\mathcal{O}$-modules, for every open set U, $H^0(U,S)$ is an $\mathcal{O}(U)$-module; if $U \subset M'$, $H^0(U,S)$ is an $\mathcal{O}(M')$-module. It is easy to dedu (using for example the Čech representation of cohomology) that the cohomology groups $H^p(M',S)$ become, in a natural way, modules over $\mathcal{O}(M')$. In other words,

$\mathcal{O}(M')$ is represented as a ring of transformations of $H^p(M',S)$. In our case, this is a finite-dimensional vector space over $\mathbb{C}$, thus there is an ideal I of finite codimension in $\mathcal{O}(M')$ such that for $f \in I$ and $\omega \in H^p(M',S)$, $f\omega = 0$. The set $|I|$ of common zeros of I is a union of level sets of $\mathcal{O}(M')$, but cannot contain infinitely many such level sets. For if $|I|$ had infinitely many level sets the restriction of $\mathcal{O}(M')$ to $|I|$ would be an infinite dimensional subspace of $C(|I|)$. This restriction is the same as $\mathcal{O}(M')/I$ on $C(|I|)$, which is finite dimensional. Thus $|I|$ must contain only finitely many level sets of $\mathcal{O}(M')$, so by the lemma $|I| \cap M$ consists of a finite union of positive dimensional compact subvarieties and finitely many isolated points, and thus is a finite union of level sets of $\mathcal{O}(M)$.

By standard local techniques (cf. p. 86 of [12] and compactness of $\overline{M}$ in M', we can find $g_1,\ldots,g_r \in I$ such that

$$V(g_1,\ldots,g_r) \cap \overline{M} = |I| \cap \overline{M} .$$

Now, our proof of theorem 3 proceeds by backward induction on p. If p is large enough, $H^p(M,S) = 0$ for <u>any</u> sheaf S [16]. Thus we may assume our theorem is true for $H^{p+1}(M,F)$, any coherent sheaf F. We shall apply this induction assumption to the kernel sheaf R of the map $G: S^r \longrightarrow S$ defined by

$$G(\sigma^1,\ldots,\sigma^r) = \sum_{i=1}^{r} g_i\sigma^i .$$

The significance of this map lies in the annihilating property of the g's. Since $H^p(M',S) \longrightarrow H^p(M,S)$ is surjective (viz. Grauert's proof of theorem 2 [8]), if $g \in I$ and $\omega \in H^p(M,S)$ we still have $g\omega = 0$. Since $g_1,\ldots,g_r \in I$, G induces the zero map on p^{th} cohomology; i.e., if $\omega \in H^p(M,S^r)$, $G\omega = 0$.

Now, by induction there are $h_1,\ldots,h_s \in \mathcal{O}(M)$ such that $V(h_1,\ldots,h_s)$ consists of finitely many level sets, and for J the idealsheaf generated by $\{h_i\}$ the restriction map $H^p(M,R) \longrightarrow H^p(M, \mathcal{O}/J \otimes_{\mathcal{O}} R)$ is injective; or, what is the same, the map $H^p(M,JR) \longrightarrow H^p(M,R)$ is the zero map. By the lemma of Artin-Rees [26], there is an n such that $R \cap J^nS^r \subset JR$. (Artin-Rees tells us this for the stalks at a point; by coherence it holds on the sheaf level in a neighborhood; by compactness of the support of J, there is an n which will do on all of M).

Let I be the idealsheaf generated by all products $\{g_i h_{i_1} \cdots h_{i_n}\}$. We enumerate these products as $f_1,\ldots,f_t$. Since

$$V(f_1,\ldots,f_t) \subset V(g_1,\ldots,g_r) \cup V(h_1,\ldots,h_s)$$

and (i) holds for the varieties on the right, it also holds for $V(f_1,\ldots,f_t)$. We have to show (ii), or what is the same, the map

$$H^p(M,IS) \longrightarrow H^p(M,S)$$

is the zero map. This amounts to an arrow-chase.

Clearly, the image of $G|J^nS^r$ is IS and the kernel is $R \cap J^nS^r$. We obtain the following commutative diagram involving cohomology:

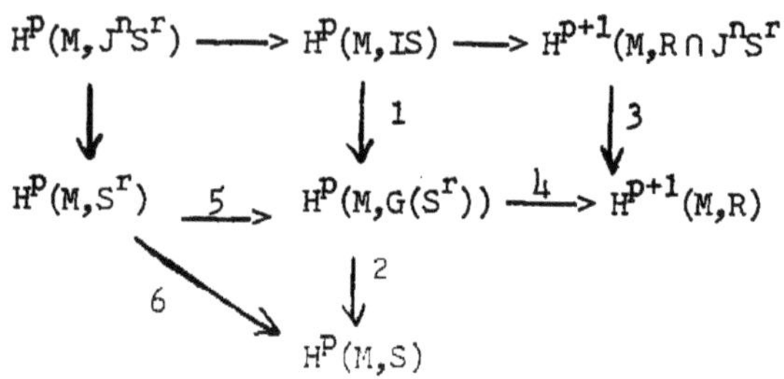

The rows are exact. We want to verify that $2 \circ 1$ is zero. Since $R \cap J^nS^r \subset JR$, 3 factors through $H^{p+1}(M, JR) \longrightarrow H^{p+1}(M,R)$, which is by the induction assumption, a zero map. Thus 3 is a zero map, so by commutativity $4 \circ 1$ is zero. By exactness $\text{Im } 1 \subset \text{Ker } 4 = \text{Im } 5$, so $\text{Im } (2 \circ 1) \subset \text{Im } (2 \circ 5) = \text{Im } 6 = 0$ as already observed. The theorem is proven.

The remark stated at the end of the theorem is clear: since $\mathcal{O}/I \otimes_{\mathcal{O}} S = 0$ off V, $H^p(M, \mathcal{O}/I \otimes_{\mathcal{O}} S) = H^p(V, \mathcal{O}/I \otimes_{\mathcal{O}} S)$. Since the pth cohomology $(p > 0)$ of any coherent sheaf vanishes at an isolated point, we may replace V by V'.

Corollary. If M is a strongly pseudoconvex domain without positive dimensional compact subvarieties, M is Stein.

Proof. In this case, applying the theorem $H^p(M,S) \longrightarrow H^p(V', \mathcal{O}/I \otimes_{\mathcal{O}} S)$ is injective, but $V' = \emptyset$. Thus $H^p(M,S) = 0$ for all coherent sheaves on X, and $p > 0$.

Corollary. Let M be a strongly pseudoconvex manifold in X. Let S be a coherent analytic sheaf on X, and $p > 0$. Suppose that $\mathfrak{U}$ is a cover of M so that all p-chains of $N(\mathfrak{U})$ are disjoint from the positive dimensional compact subvarieties of M on which the pth cohomology of S is supported. Then $H^p(N(\mathfrak{U}), S) \longrightarrow H^p(M,S)$ is the zero map.

Proof. For any compact set $K \subset M$ we have the commutative diagram

$$\begin{array}{ccc} H^p(N(\mathfrak{U}), S) & \xrightarrow{1} & H^p(N(\mathfrak{U} \cap K), S) \\ \downarrow 3 & 2 & \downarrow \\ H^p(M,S) & \longrightarrow & H^p(K,S) \end{array}$$

By the theorem and the hypotheses we can choose K so that 2 is injective and $N^p(\mathfrak{U} \cap K)$ is empty. In particular it follows that 1 is the zero map, and so $2 \cdot 3$ is zero. Since 2 is injective, 3 is zero.

Remark. In case $p = 1$, 3 is always injective, so in our case we will have $H^1(N(\mathfrak{U}), S) = 0$.

Theorem 4. Let M be a strongly pseudoconvex domain in X.

(a) if x and y belong to different connected compact subvarieties of M, there is an $f \in \mathcal{O}(M)$ such that $f(x) \neq f(y)$.

(b) if x belongs to no positive dimensional compact subvariety, there are $f_1, \ldots, f_n \in \mathcal{O}(M)$ which give local coordinates at x.

(c) M has only finitely many positive dimensional compact subvarieties.

<u>Proof</u>. (a) Let L_x, L_y be the connected compact subvarieties of M containing x,y respectively. By hypothesis $L_x \cap L_y = \emptyset$. Let I be the idealsheaf of $L_x \cup L_y$, and apply theorem 3 to I. Since the cohomology of I is supported on some finite union of compact connected varieties (which may or may not include L_x and L_y), we can find disjoint neighborhoods U_x, U_y of L_x, L_y respectively which are disjoint and contain only these relevant subvarieties. Let $U_o = M - (L_x \cup L_y$ the corollary can be applied to the covering $\mathfrak{U} = \{U_x, U_y, U_o\}$. Let $\omega \varepsilon C^o(N(\mathfrak{U}), \mathcal{O})$, $\omega(U_x) = 1$, $\omega(U_y) = 0 = \omega(U_o)$. Then $\delta\omega \varepsilon Z'(N(\mathfrak{U}), I)$, so by the corollary there is a $\theta \varepsilon C^o(N(\mathfrak{U}), I)$ such that $\delta\theta = \delta\omega$. Then $f = \theta - \omega \varepsilon \mathcal{O}(M)$ and $f(x) = 1$, $f(y) = 0$.

(b) Let m_x be the idealsheaf of functions vanishing at x. Let B be a small coordinate ball centered at x with coordinates $z_1, \ldots, z_n$. Consider the cover $\{B, M - \{x\}\}$ and the cochains $\omega_i \varepsilon C^o(N(\mathfrak{U}), m_x)$: $\omega_i(B) = z_i$, $\omega_i(M - \{x\}) = 0$. $\delta\omega_i \varepsilon Z^1(N(\mathfrak{U}), m_x^2)$, and if B is small enough, these cocycles cobound. Thus there are $\theta_i \varepsilon C^o(N(\mathfrak{U}), m_x^2)$ such that $\delta(\omega_i - \theta_i) = 0$. Let $f_i = \omega_i - \theta_i$.

(c) If ε is small enough $K = \{x \varepsilon M; \varphi(x) = -\varepsilon\}$ is a compact subset of M, disjoint from all positive dimensional subvarieties. By (b), if $x \varepsilon K$, there are $f_1^x, \ldots, f_n^x \varepsilon \mathcal{O}(M)$ which are coordinates in a neighborhood U_x of x. Cover K by finitely many such neighborhoods, and let $\{f_1,\dots,f_s\}$ be the union of all the associated functions. Let J be the set of all $x \in M$ at which the map $F=(f_1,\dots,f_s)$ is singular (i.e., $df_1,\dots,df_s$ do not span m_x/m_x^2). J is a subvariety of M, and $J \cap K = \emptyset$. Thus $J_0 = J \cap \{x \in M; \varphi(x) < -\varepsilon\}$ is a compact subvariety of M. Now if $x \notin J$, the functions $f_1,\dots,f_s$ separate points in some neighborhood of x, so x is not in any positive dimensional level set. Thus the positive dimensional connected components of J are precisely all the positive dimensional level sets. Since J is compact, they are finite in number.

For M a strongly pseudoconvex manifold, the union E of all the positive dimensional compact subvarieties of M is called the <u>exceptional set</u> of M. If we identify each connected component of E to a point we obtain a topological space $\tilde{M}$ and a proper map $\pi: M \longrightarrow \tilde{M}$. We can make $\tilde{M}$ into a ringed space by defining the sheaf of functions $\tilde{\mathcal{O}}$ as $\tilde{\mathcal{O}}(U) = \mathcal{O}(\pi^{-1}(U))$. It is easy to see (via the proper mapping theorem) that this ringed space is an analytic space. In fact, it is a Stein analytic space, because it is again strongly pseudoconvex with <u>no</u> positive dimensional compact subvarieties. (the presence of the isolated singularities does not disturb any of the above arguments). Thus

<u>Theorem 5</u>. A strongly pseudoconvex manifold is a proper modification of a Stein analytic space with isolated singular points.

Conversely, by Hironaka's resolution of singularities [11], there is always a proper modification of some neighborhood of an isolated singularity which is a strongly pseudoconvex manifold. This fact allows the possibility of studying isolated singular points by pseudoconvexity methods ([11,13,34]).

4. Partial Differential Equations.

Since the Cauchy-Riemann equations serve to characterize holomorphic functions, the problems discussed above can be viewed as part of the study of this system of partial differential equations. During the 1950's this was a fruitful approach in the study of compact manifolds; in the past decade, with the solution of certain boundary value problems this approach has been broadened to include pseudoconvex manifolds. In fact Morrey [25] first proved that certain pseudoconvex manifolds were holomorphically convex by means of L^2 techniques in partial differential equations. J. J. Kohn [17] developed a general technique for representing the Dolbeault cohomology groups by harmonic forms and gave a new proof of theorem 2 (for locally free sheaves). This proof (and even more so, subsequent work of Hörmander [14,15]) implies the existence of estimates on a natural Hilbert space structure on the vector spaces associated with cohomology theory. For example, an important typical result is this. Suppose D is a relatively compact pseudoconvex domain in $\mathbb{C}^n$. We consider the norm $||\cdot||$ on forms induced by the Euclidean metric and Lebesque measure on D.

Theorem 6. (Hörmander). If ω is an L^2 (p,q) form on D $(q>0)$ such that $\bar{\partial}\omega=0$, then $\omega=\bar{\partial}\theta$ where θ is an L^2 $(p,q-1)$ form on D. Further, there is a constant K depending only on diam D such that θ can be chosen with $||\theta|| \leq K||\omega||$.

This estimate could not be obtained by the Čech cohomological methods. From it come (for example) approximation theorems (cf. [14], [29]) like the following:

<u>Theorem 7</u>. (Nirenberg-Wells). Let M be a real submanifold of $\mathbb{C}^n$ with this property: for all $x \in M$, if T_x is the vector space tangent to M at x, $T_x \cap iT_x = \{0\}$. Then every continuous function on M can be uniformly approximated by functions holomorphic in a neighborhood of M.

Finally, one can study a strongly pseudoconvex domain by studying only its boundary, for the boundary determines the function-theory of the domain [32]. Kohn has developed the necessary theory to apply these same methods to the boundary of a pseudoconvex domain [21].

Bibliography

1. Andreotti, A., and Grauert, H., "Théorémes de finitude pour la cohomologie des espaces complexes," Bull. Soc. Math. France 90 (1962), 193-259.

2. Andreotti, A., "Théorémes dé dependance algébrique sur les espaces complexes pseudo-concaves," Bull. Soc. Math. France 91 (1963), 1-38.

3. Bremermann, H. J., Über die Äquivalenz der pseudo-konvexen Gebiete und der Holomorphie-gebiete im Raum von n komplexen Veränderlichen," Math. Ann. 128 (1954), 63-91.

4. Bremermann, H. J., "Complex convexity," Trans. Amer. Math. Soc. 82 (1956) 17-51.

5. Cartan, H., Séminaire E.N.S., 1951-1952, École Normale Supérieure, Paris.

6. Docquier, F., and Grauert, H., "Levisches Problem und Rungescher Satz für Teilgebiete Steinscher Mannigfaltigkeiten," Math. Ann. 140 (1960), 94-123.

7. Garabedian, P. R., and Spencer, D. C., "Complex boundary value problems," Trans. Amer. Math. Soc. 73 (1952), 223-242.

8. Grauert, H., "On Levi's Problem and the imbedding of real-analytic manifolds," Ann. Math. 68 (1958), 460-472.

9. Grauert, H., "Über Modifikationen und exzeptionelle analytische Mengen," Math. Ann. 146 (1962), 331-368.

10. Grauert, H., und Remmert, R., "Plurisubharmonische Funktionen in Komplexen Räumen," Math. Zeit. 65 (1956), 175-194.

11. Hironaka, H., "The resolution of singularities of an algebraic variety (characteristic zero)," Ann. Math. 79 (1964), 109-800.

12. Gunning, R. C. and Rossi, H., Analytic Functions of Several Complex Variables, Prentice-Hall, 1965.

13. Hironaka, H., A Fundamental lemma on point modifications, Conference on Complex Analysis, Minneapolis, Springer-Verlag 1965.

14. Hörmander, L., L^2 estimates and existence theorems for the $\bar{\partial}$-operator, Acta Math. 113, 89-152 (1965).

15. Hörmander, L., An Introduction to Complex Analysis in Several Variables, Van Nostrand, 1966.

16. Hurewicz and Wallman, Dimension theory.

17. Kohn, J. J., "Harmonic integrals on strongly pseudoconvex manifolds, I.,II," Ann. Math. 78 (1963), 112-148.

18. Kohn, J. J. and Nirenberg, L., Non-coercive boundary problems, Comm. Pure and Appl. Math. 18, 443-492 (1965).

19. Kohn, J. J., and Rossi, H., "On the extension of holomorphic functions from the boundary of a complex manifold" (not published).

20. Kohn, J. J., and Spencer, D. C., "Complex Neumann problems," Ann. Math. 66 (1957), 89-140.

21. Kohn, J. J., Boundaries of complex manifolds, Conference on Complex Analysis, Minneapolis, Springer-Verlag 1965.

22. Lelong, P., "La convexité et les fonctions analytiques de plusieur variables complexes," J. Math. Pures Appl. 31 (1952), 191-219.

23. Lelong, P., "Domaines convexes par rapport aux fonctions plurisousharmoniques," J. d'Analyse Math. 2 (1952-53), 178-208.

24. Levi, E. E., "Studii sui punti singolari essenziali delle funzioni analitiche di due o più variabili complesse, "Annali di Mat. pura ed appl. 17, 3(1910), 61-87.

25. Morrey, C. B., "The analytic imbedding of abstract real-analytic manifolds," Ann. Math. 68 (1958), 159-201.

26. Nagata, M., Local Rings, Interscience 1962.

27. Narasimhan, R., "The Levi problem for complex spaces," Math. Ann. 142 (1961), 355-65.

28. Narasimhan, R., "Levi Problem for Complex Spaces II," Math. Ann. 146 (1962), 195-216.

29. Nirenberg, R. and Wells, R. O., Holomorphic approximation on real submanifolds of complex manifolds, Bull. A.M.S. 73, 378-381 (1967).

30. Norguet, F., "Sur les domaines d'holomorphie des fonctions uniformes de plusieurs variables complexes (passage du local au global)," Bull. Soc. Math. France 82 (1954), 137-159.

31. Oka, K., Sur les fonctions analytiques de plusieurs variables (Tokyo, Iwanami Shoten, 1961). [This is a collection of reprints of nine articles under the same general title, which have appeared in the following journals:

I "Domaines convexes par rapport aux fonctions rationelles," J. Sci. Hiroshima Univ., ser. A 6 (1936), 245-255.

II "Domaines d'holomorphie," J. Sci. Hiroshima Univ., ser. A 7 (1937), 115-130.

III "Deuxième|problème de Cousin," J. Sci. Hiroshima Univ., ser. A 9 (1939), 7-19.

IV "Domaines d'holomorphie et domaines rationellement convexes," Jap. J. Math. 17 (1941), 517-521.

V "L'intégrale de Cauchy," Jap. J. Math. 17 (1941), 523-531.

VI "Domaines pseudoconvexes," Tohoku Math. J. 49 (1942), 15-52.

VII "Sur quelques notions arithmétiques," Bull. Soc. Math. France 78 (1950), 1-27.

VIII "Lemme fondamental," J. Math. Soc. Japan 3 (1951), 204-214 and 259-278.

IX "Domaines finis sans point critique intérieur," Jap. J. Math. 23 (1953), 97-155.

Since then, the following paper in the series has also appeared:

X "Une mode nouvelle engendrant les domaines pseudoconvexes," Jap. J. Math. 32 (1962), 1-12.]

32. Rossi, H., Attaching analytic spaces to an analytic space along a pseudoconcave boundary, Conference on Complex Analysis, Minneapolis, Springer-Verlag, 1965.

33. Rossi, H., Lecture notes, Seminaires des Mathematiques Supérieures, Montréal, 1967.

34. Hironaka, H. and Rossi, H., On the equivalence of embeddings of exceptional complex spaces, Math. Annalen, 1965.

BANACH ALGEBRAS AND UNIFORM APPROXIMATION.

John Wermer

1. INTRODUCTION:

Let $\mathbb{C}^n$ be the space of n complex variables and consider a compact subset X of $\mathbb{C}^n$.

We denote by P(X) the space of all functions defined on X which are uniform limits of polynomials in the coordinates $z_1,\ldots,z_n$ C(X) is the space of all continuous complex-valued functions on X. Our main problem is to describe the functions that lie in P(X) . When n = 1, i.e. X is a plane set, the complete answer is known, but for $n > 1$ the problem is solved only in certain special cases.

P(X) is a closed subalgebra of C(X) , hence a Banach algebra, evidently commutative, semi-simple and with unit. What is its Gelfand representation?

We define the hull of X , written h(X) , as the set of all z in $\mathbb{C}^n$ such that for every polynomial Q

$$|Q(z)| \leq \max_X |Q| \quad .$$

Each point z^o of h(X) defines, by evaluation, a homomorphism of the algebra of polynomials into the scalars which extends to a homomorphism of P(X) and so yields a point in the maximal ideal space of P(X) . It is easily seen then that we can identify h(X) with the maximal ideal space of P(X) . Also, the Gelfand image $\hat{f}$ of an element f of P(X) can be found as follows: choose z in h(X) and an arbitrary sequence $\{P_n\}$ of polynomials tending to f on X . Then

$$\hat{f}(z) = \lim_n P_n(z) \quad .$$

Of course $h(X)$ contains X and $\hat{f}$ coincides with f on X.

A set X such that $h(X) = X$ is called polynomially convex. Since $C(X)$ has X as its maximal ideal space, a necessary condition on X in order that every continuous function on X be uniformly approximable by polynomials, i.e. that $P(X) = C(X)$, is that X be polynomially convex. This raises two questions:

A. When is a given set X in $\mathbb{C}^n$ polynomially convex? and

B. What conditions beyond polynomial convexity are sufficient to give $P(X) = C(X)$?

I shall discuss these questions for $n>1$ when X is a submanifold of $\mathbb{C}^n$. But first let us look at the case: $n = 1$.

2. MERGELYAN'S THEOREM.

X here is a subset of the complex plane. $h(X)$ is seen to be the union of X and its bounded complementary components. For f in $P(X)$ and z in $h(X) - X$, $\hat{f}(z)$ is the value at z of the analytic extension of f to the component containing z. In studying $P(X)$ we may without loss of generality assume

(1) X is the boundary of its unbounded complementary component.

Mergelyan's Theorem: Assume (1). Then $P(X)$ consists of all functions in $C(X)$ which admit analytic extension to the interior of $h(X)$.

Mergelyan proved this theorem using honest function theory [1]. How would a functional analyst try to prove it? He would try to find all continuous linear functionals on $C(X)$ which vanish on $P(X)$.

We may look at this problem more generally. Let Y be a compact space and A a separating algebra of continuous functions on Y which contains the constants. How can we get linear functionals on $C(Y)$ vanishing on A ?

Let us choose a maximal ideal m of A and a representing measure σ for m , i.e. a probability measure σ on Y with

$$\int f \, d\sigma = \hat{f}(m) \quad \text{for all } f \text{ in } A.$$

Define $H^1(\sigma)$ as the closure of A in $L^1(\sigma)$. Choose k in $H^1(\sigma)$ such that

$$\int k \, d\sigma = 0.$$

Then for all f in A,

$$\int f \cdot k \, d\sigma = \int f \, d\sigma \cdot \int k \, d\sigma = 0,$$

and so $k\,d\sigma$ provides a linear functional on $C(Y)$ vanishing on A. Provided that σ is not a point mass, we can choose k so that $k\,d\sigma$ is not zero. We call such a functional $k\,d\sigma$ elementary. By taking different m, σ, k and forming linear combinations of the corresponding elementary functionals we can build up linear functionals on $C(Y)$ which annihilate A.

<u>Theorem</u> 2.1 : <u>Let</u> X <u>satisfy</u> (1). <u>Then every functional which annihilates</u> $P(X)$ <u>is a linear combination, possibly infinite, of elementary functionals. More precisely:</u>

<u>Let</u> $\{\Omega_i\}$ <u>be the bounded complementary components of</u> X. <u>In</u>

each Ω_i choose a point a_i and a representing measure σ_i for a_i. Let μ be an arbitrary finite complex measure on X annihilating P(X). Then we can find elements k_i in $H^1(\sigma_i)$ with $\int k_i \, d\sigma_i = 0$ so that

(2) $$\mu = \sum_i k_i \, d\sigma_i \ ,$$

the series converging in total variation norm.

Mergelyan's theorem readily follows from (2). For this deduction and for the proof of Theorem 2.1 see [2], [3], [4]. The ideas here go back to work of Bishop in [5], [6], [7]. Let us note two consequences of Theorem 2.1.

Corollary 1 : Let X be the unit circumference. Then there is only one Ω_i, the open disk, and we can take $a_1 = 0$ and $\sigma_1 = \frac{1}{2\pi} d\theta$ We thus have that if μ is a measure on $|z| = 1$ with

$$\int z^n \, d\mu = 0 \ , \ n \geqslant 0,$$

then there is some k in $H^1(d\theta)$ so that $\mu = k \, d\theta$. But $H^1(d\theta)$ is the usual Hardy class H^1, on the boundary, and so we have recovered the theorem of F. and M. Riesz. (See [8]).

Corollary 2 : Assume that the complement of X is connected. Then the sum in (2) is empty and so the only functional annihilating P(X) is zero, whence P(X) = C(X).

This is of course a special case of Mergelyan's theorem, originally due to Lavrentiev.

Mergelyan has extended his result in various ways to approximation by rational functions [9]. Let E be a compact plane set. Necessary and sufficient conditions on E that every continuous function on E which is analytic at the interior points be a uniform limit on E of rational function with poles outside E have recently given by Vituškin [10].

A number of results on rational approximation have also been obtained by means of functional analysis [11], [12], [13], [14].

For $\mathbb{C}^n$ with $n > 1$ we know almost nothing about the measures on a set X which annihilate P(X). A formula of the form (2) is not possible in general.

3. HIGH DIMENSIONAL MANIFOLDS IN $\mathbb{C}^n$.

If X is a simple closed curve in the plane, X cannot be polynomially convex. There is an analogue of this in $\mathbb{C}^n$ (we recall that the topological dimension of $\mathbb{C}^n$ is 2n).

Theorem 3.1 : *A homeomorphic image X in $\mathbb{C}^n$ of an n-dimensional sphere is not polynomially convex.*

This depends on a result of Serre [15] and is proved by A. Browder [16].

Corollary : *Let X be a compact k-dimensional manifold in $\mathbb{C}^n$, with or without boundary. Assume $k \geq n$. Then $P(X) \neq C(X)$.*

For X contains a topological n-sphere X_1 . By the theorem, $h(X_1) \neq X_1$. Hence $P(X_1) \neq C(X_1)$. Since each element of $C(X_1)$ extends to an element of C(X) , it follows that $P(X) \neq C(X)$.

For smooth k-manifolds X in $\mathbb{C}^n$, $k > n$, satisfying certain restrictions, one can show that $h(X)$ has analytic structure, i.e. contains pieces of complex analytic manifolds. [17], [18], [19].

4. ONE DIMENSIONAL MANIFOLDS IN $\mathbb{C}^n$.

If J is a homeomorphic image of the unit interval lying in the complex plane, then $P(J) = C(J)$ by Corollary 2 of Theorem 2.1. This was first shown by Walsh [20].

In $\mathbb{C}^n$, $n > 1$, the corresponding assertion if false. [21], [22]. There exist Jordan arcs J which fail to be polynomially convex and so with $P(J) \neq C(J)$. However if we assume in addition that J is smooth (continuously differentiable), all is well. We have

Theorem 4.1 : *J is polynomially convex and $P(J) = C(J)$.*

The hard thing to show here is that a smooth arc is polynomially convex. A simple, direct proof of this fact would be very interesting. For Γ a compact 1-dimensional manifold in $\mathbb{C}^n$, i.e. a simple closed curve, we have in addition:

Theorem 4.2 : *If Γ is smooth, then either Γ is polynomially convex or $h(\Gamma) - \Gamma$ is a one-complex-dimensional analytic space.*

Theorems 4.1 and 4.2 were proved in the real-analytic case by the author. [23]. In the smooth case they were proved (unpublished) by Bishop as consequence of his work in [24], and by another method, as consequence of his paper [25]. They were proved and published by Stolzenberg in [26], in more general form.

A related problem which despite much effort is still unsolved is this: Let J be a homeomorphic image of the unit interval in $\mathbb{C}^n$. Assume that J is polynomially convex. Does it follow (without smoothness assumptions) that $P(J) = C(J)$? For smooth J this implication was proved in [27].

5. MANIFOLDS WITH NO COMPLEX TANGENTS.

Let S be a smooth manifold in $\mathbb{C}^n$ which contains a complex-analytic manifold S^o. Let X be a compact subset of S. Suppose now that X contains a relatively open set N in S^o. Then every function in $P(X)$ will be analytic when restricted to N, and so $P(X) \neq C(X)$.

Fix x in S and let T_x denote the tangent space to S at x, viewed as a real subspace of $\mathbb{C}^n$. By a complex tangent to S at x we mean a complex line (one-dimensional complex linear subspace) contained in T_x. Let S^o be as above and fix x in S^o. Since S^o evidently possesses a complex tangent at x, it follows that S has a complex tangent at x.

To rule out the possibility that S contains complex-analytic submanifolds we can thus require that S have no complex tangents. It turns out that this condition is just what is needed to answer question B of the Introduction, for subsets X of S.

In the case when S is a smooth 2-dimensional disk in $\mathbb{C}^2$ the author showed the following [28]: Let f be a complex-valued function defined and of class C^1 in an open neighborhood N of the closed unit disk in the z-plane. Let S be the image in $\mathbb{C}^2$ of N under the map: $z \to (z, f(z))$ and let X be the image of the closed unit disk under this map. Then

Theorem 5.1: *If* S *has no complex tangents and if* X *is polynomially convex, then* $P(X) = C(X)$.

The absence of complex tangents here amounts to the non-vanishing of $\frac{\partial f}{\partial \bar{z}}$.

This result was extended to general smooth 2-manifolds in $\mathbb{C}^n$ by Freeman in [29]. Similar results for manifolds of dimensions $k > 2$ have recently been obtained, based on the solution of the so-called "$\bar{\partial}$-Neumann problem" by Morrey [30], Kohn [31], and Hörmander [32].

Consider a domain Ω in $\mathbb{C}^n$ and a form f of type (0,1) defined in Ω : $f = \sum_{j=1}^{n} f_j \, d\bar{z}_j$. We seek a function u in Ω satisfying

$$\bar{\partial} u = f \,, \quad \text{i.e.} \; \frac{\partial u}{\partial \bar{z}_j} = f_j, \; j = 1,2,\ldots,n. \tag{3}$$

A necessary condition for (3) is

$$\bar{\partial} f = 0, \; \text{i.e.} \; \frac{\partial f_j}{\partial \bar{z}_k} = \frac{\partial f_k}{\partial \bar{z}_j}, \; j,k = 1,\ldots,n. \tag{4}$$

The $\bar{\partial}$-Neumann problem (in the form needed here) is to show that for a certain class of domains Ω (the pseudo-convex domains) if (4) holds, then (3) has a solution. For a discussion of this problem see e.g. Hörmander's book [33], Chapter IV.

Nirenberg and Wells have announced a number of approximation theorems for k-dimensional smooth manifolds with arbitrary k [34]. In particular

Theorem 5.2: *Let* M *be a compact* C^∞ *submanifold of* $\mathbb{C}^n$ *having no complex tangents. Then every continuous function on* M *is a uniform limit of functions holomorphic in a neighborhood of* M .

The proof in [34] is based on Kohn's work in [31]. Hörmander and the author, making use of results of [32], have proved some related results [35]. In particular

Theorem 5.3 : *Let* S *be a* k*-dimensional submanifold of* $\mathbb{C}^n$ *which is of class* C^r *, where* $r \geq \frac{k}{2} + 1$, *such that* S *has no complex tangents. If* X *is a compact polynomially convex subset of* S *, then* $P(X) = C(X)$.

6. PERTURBATION THEOREMS.

Fix a compact set X in $\mathbb{C}^n$ and continuous functions $f_1,\ldots,f_k$ defined on X . Denote by $[f_1,\ldots,f_k;X]$ the class of all functions on X which are uniform limits of polynomials in $f_1,\ldots,f_k$. The Stone-Weierstrass theorem tells us that

$$[z_1,\ldots,z_n,\bar{z}_1,\ldots,\bar{z}_n;\ X] \;=\; C(X).$$

Suppose now that g_j is a function "near" $\bar{z}_j$ on X for $j = 1,\ldots,n$. Does it follow that

$$[z_1,\ldots,z_n,g_1,\ldots,g_n;\ X] \;=\; C(X)\ ?$$

If "near" is taken to mean uniformly close on X , the answer is No. For we may for instance take X to be the closed unit disk in the plane. Fix $\varepsilon > 0$. Then we can choose g in $C(X)$ with $g(z) = 0$ in $|z| < \varepsilon$ and $|g(z) - \bar{z}| \leq \varepsilon$ on X. Evidently

$$[z,g;\ X] \neq C(X),$$

since every element of $[z,g;X]$ is analytic in $|z| < \varepsilon$.

However, if we interpret "near" as closeness in the Lipschitz norm instead, the answer is Yes. In [36], the author showed

Theorem 6.1 : *Denote by* D *the closed unit disk in the plane. Put* $g(z) = \bar{z} + R(z)$, *where we assume*

$$|R(z) - R(a)| < |z - a|$$

for all a,z *in* D *with* $a \neq z$. *Then* $[z,g;D] = C(D)$.

In [35] Hörmander and the author give the following generalization of this result, obtained as a Corollary of Theorem 5.3 stated in the preceding Section:

Theorem 6.2 : *Let* X *be a compact set in* $\mathbb{C}^n$ *and let* N *be a neighborhood of* X. *Consider a vector function* $R = (R_1,\dots,R_n)$ *with values in* $\mathbb{C}^n$, *defined and of class* C^{n+1} *in* N. *Suppose that there is a constant* $k < 1$ *such that*

$$(5) \qquad |R(z) - R(z')| \leq k|z - z'| \quad \text{for all } z,z' \text{ in } N.$$

Then $[z_1,\dots,z_n, \bar{z}_1 + R_1,\dots,\bar{z}_n + R_n;X] = C(X)$.

If we replace k in (5) by a constant ≥ 1, the conclusion may fail to be true. Take, e.g., $R(z) = -\bar{z}$, and X to be a set with interior.

REFERENCES.

1. S. N. Mergelyan, On the representation of functions by series of polynomials on closed sets, *Doklady Akad. Nauk. SSSR*, (N.S.) 78 (1951), 405-408, (Russian)
2. I. Glicksberg and J. Wermer, Measures orthogonal to a dirichlet algebra, *Duke Math. Jour.* 30 (1963), 661-666
3. J. Wermer, Seminar über Funktionen - Algebren, *Lecture Notes in Mathematics* 1, Springer Verlag, (1964)
4. L. Carleson, Mergelyan's theorem on uniform polynomial approximation, *Math. Scand.* 15 (1964), 167-175
5. E. Bishop, The structure of certain measures, *Duke Math. Jour.* 25 (1958), 283-289
6. E. Bishop, Boundary measures of analytic differentials, *Duke Math. Jour.* 27 (1960), 331-340
7. E. Bishop, A minimal boundary for function algebras, *Pacific Jour. of Math.* 9 (1959), 629-642
8. H. Helson and D. Lowdenslager, Prediction theory and Fourier series in several variables, *Acta Math.* 99 (1958), 165-202
9. S. N. Mergelyan, Uniform approximation to functions of a complex variable, *Amer. Math. Soc. Transl.* 101 (1954)
10. A. G. Vituškin, Necessary and sufficient conditions on a set in order that any continuous function analytic at the interior points of the set may admit uniform approximation by rational fractions, *Soviet Math. Dokl.* 7 (1966), 1622-1625

11. I. Glicksberg, Dominant representing measures and rational approximation, to appear in *Trans. Amer. Math. Soc.*

12. I. Glicksberg, The abstract F. and M. Riesz theorem, *Jour. of Funct. Anal.* 1 (1967), 109-122

13. P. R. Ahern and D. Sarason, On some hypo-dirichlet algebras of analytic functions, to appear in *Amer. Jour. of Math.*

14. J. Garnett, On a theorem of Mergelyan, to appear

15. J.-P. Serre, Une propriété topologique des domaines de Runge, *Proc. Amer. Math. Soc.* 6 (1955), 133-134

16. A. Browder, Cohomology of maximal ideal spaces, *Bull. Amer. Math. Soc.* 67 (1961), 515-516

17. H. Lewy, On the local character of the solutions of an atypical linear differential equation in three variables and a related theorem for regular functions of two complex variables, *Ann. of Math.* 64 (1956), 514-522

18. E. Bishop, Differentiable manifolds in complex Euclidean space, *Duke Math. Jour.* 32 (1965), 1-22

19. R. O. Wells, Jr., On the local holomorphic hull of a real submanifold in several complex variables, *Comm. on Pure and Appl. Math.* 19 (1966), 145-165

20. J. L. Walsh, Über die Entwicklung einer Funktion einer komplexen Veränderlichen nach Polynomen, *Math. Ann.* 96 (1926), 437-450

21. J. Wermer, Polynomial approximation on an arc in $\mathbb{C}^3$, *Ann. Math.* 62 (1955), 269-270

22. W. Rudin, Subalgebras of spaces of continuous functions, *Proc. Amer. Math. Soc.* 7 (1956), 825-830

23. J. Wermer, The hull of a curve in C^n, *Ann. of Math.* 68 (1958), 550-561

24. E. Bishop, Analyticity in certain Banach algebras, *Trans. Amer. Math. Soc.* 102 (1962), 507-544

25. E. Bishop, Holomorphic completions, analytic continuations and the interpolation of semi-norms, *Ann. of Math.* 78 (1963), 468-500

26. G. Stolzenberg, Uniform approximation on smooth curves, *Acta Math.* 115 (1966), 185-198

27. H. Helson and F. Quigley, Existence of maximal ideals in algebras of continuous functions, *Proc. Amer. Math. Soc.* 8 (1957), 115-119

28. J. Wermer, Polynomially convex disks, *Math. Ann.* 158 (1965), 6-10

29. M. Freeman, Some conditions for uniform approximation on a manifold, *Function Algebras,* (Proceedings of a Symposium), Scott Foresman and Co. (1965), 42-60

30. C. B. Morrey, The analytic embedding of abstract real analytic manifolds, *Ann. Math.* 68 (1958), 159-201

31. J. J. Kohn, Harmonic Integrals on strongly pseudoconvex manifolds, I, II, *Ann. Math.* 78 (1963), 112-148 and 79 (1963), 450-472

32. L. Hörmander, L^2 estimates and existence theorems for the $\bar{\partial}$-operator, *Acta Math.* 113 (1965), 89-152

33. L. Hörmander, *An introduction to complex analysis in several variables,* D. Van Nostrand Co. (1966)

34. R. Nirenberg and R. O. Wells, Jr., Holomorphic approximation on real submanifolds of a complex manifold, *Bull. Amer. Math. Soc.* 73, (1967), 378-381

35. L. Hörmander and J. Wermer, Uniform approximation on compact sets in C^n , (to appear)

36. J. Wermer, Approximation on a disk, Math. Ann. 155 (1964), 331-333

Extension of results from several complex variables to general function algebras.

By C. E. Rickart*

Introduction.

The remarks which follow are concerned with the problem of obtaining, for general function algebras, properties that have an analytic character. Although most of the research on function algebras is motivated in one way or another by questions of analyticity, the bulk of such work has been devoted to the extension of classical analytic results to certain special algebras. The purpose here is to present certain analytic type results that hold for essentially arbitrary function algebras. The extent to which we must qualify the work "arbitrary" will become clear later. In any case, we do not wish to exclude examples of the type constructed by Stolzenberg [13] and discussed by Gleason in his lecture in this series. Therefore one cannot in general ask for a relevant analytic structure in the spaces of maximal ideals of such algebras. Nevertheless, there remain properties, which may legitimately be called "analytic", that are shared by virtually all function algebras. Perhaps the best example of such a property is the Rossi local maximum modulus principle [12]. Although this is a "weak" maximum principle, it is genuinely analytic in character and holds for any uniform algebra $\mathfrak{A}$ on its space Ω of maximal ideals. A convenient form of the principle asserts that if U is an open set in Ω and $\partial_{\mathfrak{A}}\Omega$ is the Šilov boundary of Ω with respect to $\mathfrak{A}$, then the maximum modulus of each function in $\mathfrak{A}$ on $\bar{U}$, the closure of U, is attained on the set $(\text{bdry } U) \cup (\bar{U} \cap \partial_{\mathfrak{A}}\Omega)$, where bdry U is the topological boundary of U. The proof, which involves the solution of a Cousin problem in $\mathbb{C}^n$, depends in an essential way on the condition that Ω be the space of maximal ideals of $\mathfrak{A}$. This condition, which is required for most of our discussion,

---

*The results outlined here were obtained under partial support from research grants AFOSR-407-63 and NSF-GP-5493.

represents virtually the only restriction on the generality of the algebras that we consider. Observe, however, that the above local maximum principle is interesting only for those algebras for which the Šilov boundary does not exhaust the space of maximal ideals.

Although function algebras are usually defined on a compact Hausdorff space, it turns out to be convenient for us to start with a more general situation without any compactness restrictions whatsoever.

§1. Natural systems.

Let Σ be an arbitrary Hausdorff space (not assumed to be compact, or even locally compact) and let $\mathfrak{A}$ denote an algebra of continuous functions on Σ that contains the constants. Note that $\mathfrak{A}$ may contain unbounded functions. The following conditions, which are imposed on the pair $[\Sigma, \mathfrak{A}]$, enable us to obtain results of an analytic nature for $\mathfrak{A}$:

I. The topology of Σ is the weakest (or coarsest) with respect to which each function in $\mathfrak{A}$ is continuous.

II. Every continuous homomorphism of $\mathfrak{A}$ onto $\mathbb{C}$ is given by evaluation at a point of Σ. (Where the topology in $\mathfrak{A}$ is uniform convergence on compact subsets of Σ.)

When conditions I and II are satisfied, we call the pair $[\Sigma, \mathfrak{A}]$ a natural system [10, Def. 1.1]. Condition II is the crucial one and is the source of the main properties that we have been able to establish for these algebras. We shall assume throughout that $[\Sigma, \mathfrak{A}]$ is natural.

The following are some of the important examples of natural systems:

1. $[\Phi_{\mathfrak{B}}, \hat{\mathfrak{B}}]$, where $\mathfrak{B}$ is a commutative Banach algebra with 1, $\Phi_{\mathfrak{B}}$ is its space of maximal ideals, and $\hat{\mathfrak{B}}$ is the Gelfand representation of $\mathfrak{B}$ on $\Phi_{\mathfrak{B}}$.

2. $[\mathbb{C}^n, \mathcal{P}]$, where $\mathbb{C}^n$ is complex n-space and $\mathcal{P}$ is the algebra of all polynomials in n complex variables.

3. $[\mathbb{C}^{\Lambda}, \mathcal{P}]$, where Λ is an arbitrary index set, $\mathbb{C}^{\Lambda}$ is the cartesian product of "Λ" complex planes, $\{\mathbb{C}_\lambda : \lambda \in \Lambda\}$, and $\mathcal{P}$ is the algebra of all ordinary polynomials in the complex variables $\{\zeta_\lambda : \lambda \in \Lambda\}$

4. $[\mathcal{V}, \mathcal{O}_{\mathcal{V}}]$, where $\mathcal{V}$ is a Stein manifold and $\mathcal{O}_{\mathcal{V}}$ is the algebra of all holomorphic functions on $\mathcal{V}$.

Although example 1 is very important, it is example 2 that provides the main guide for our investigation. In fact, at the present stage, the general program is to extend certain results for analytic functions in $\mathbb{C}^n$ to arbitrary natural systems.

A fundamental concept for $[\Sigma, \mathfrak{A}]$ is a generalization of the notion of polynomial convexity. Let K be a compact subset of Σ. Then the set

$$\hat{K} = \{\sigma : \sigma \in \Sigma, \ |a(\sigma)| \leq |a|_K, \ a \in \mathfrak{A}\},$$

where $|a|_K$ denotes the supremum of $|a(\sigma)|$ on K, is called the

$\mathfrak{A}$ - <u>convex hull</u> of K. It is always true that $K \subseteq \hat{K}$ and, if $K = \hat{K}$, the set K is said to be $\mathfrak{A}$ -<u>convex</u>. An important consequence of the naturality is that the $\mathfrak{A}$ -convex hull of a compact set is also compact [8]. An arbitrary set Ω in Σ is said to be $\mathfrak{A}$ -<u>convex</u> if for every compact set $K \subseteq \Omega$ it is true that $\hat{K} \subseteq \Omega$. Let $[\Omega , \mathfrak{A}]$ denote the pair $[\Omega , \mathfrak{A}/\Omega]$, where $\mathfrak{A}/\Omega$ is the restriction of $\mathfrak{A}$ to the set Ω . Then another basic result is that $[\Omega , \mathfrak{A}]$ is natural if and only if Ω is $\mathfrak{A}$ -convex [10, Prop. 1.3]. It is through this "localization" to compact $\mathfrak{A}$ -convex sets that results for the compact case become available in the non-compact case. In particular, we are able to use the Rossi local maximum modulus principle in this way. It is mainly through the maximum principle, which turns out to be a powerful tool, that our study of general algebras depends on several complex variables.

§2. $\mathfrak{A}$ -holomorphic functions.

We now define a family of functions in Σ which corresponds to the family of ordinary holomorphic functions in $\mathbb{C}^n$. It is convenient to start with some preliminary definitions.

Let $\mathcal{F}$ be a given family of functions defined on subsets of Σ . A function g defined on a set $X \subseteq \Sigma$ is said to be locally approximable by elements of $\mathcal{F}$ if for $\zeta \in X$ there exists a neighborhood N_ζ of ζ such that, on the set $X \cap N_\zeta$, g is a uniform limit of functions defined on $X \cap N_\zeta$ and belonging to $\mathcal{F}$. The collection of all functions that are locally approximable by elements of $\mathcal{F}$ is called the local extension of $\mathcal{F}$ and denoted by loc $\mathcal{F}$. If $\mathcal{F}$ = loc $\mathcal{F}$, then $\mathcal{F}$ is said to be locally closed. Note that, if $\mathcal{F}$ contains a certain function, then loc $\mathcal{F}$ contains every function obtained by restricting the given function to an arbitrary subset of its domain of definition. The family of all functions defined on subsets of Σ is obviously locally closed. Moreover, the intersection of any collection of locally closed families is again locally closed. Therefore the smallest locally closed family that contains a given $\mathcal{F}$ exists. We call it the local closure of $\mathcal{F}$ and denote it by $\mathcal{F}_{loc}$ [10, §2].

For the natural system $[\Sigma, \mathfrak{A}]$, we define as $\mathfrak{A}$ -holomorphic those functions that belong to the local closure, $\mathfrak{A}_{\text{loc}}$, of the algebra $\mathfrak{A}$. There is a convenient transfinite decomposition of $\mathfrak{A}_{\text{loc}}$ which we now describe. An ordinal μ exists such that to each $\nu \leq \mu$ there corresponds a family $\mathfrak{A}_\nu$ of functions satisfying the following conditions:

(1) $\mathfrak{A}_0 = \mathfrak{A}$, $\mathfrak{A}_\mu = \mathfrak{A}_{\text{loc}}$, and
$\mathfrak{A}_\alpha \subsetneq \mathfrak{A}_\beta$ for $\alpha < \beta \leq \mu$.

(2) For each $\nu \leq \mu$, $\mathfrak{A}_\nu = \text{loc}\,(\bigcup_{\alpha<\nu} \mathfrak{A}_\alpha)$

Elements of $\mathfrak{A}_\nu$ are said to be $\mathfrak{A}$ -holomorphic of class ν [10].

Using this decomposition, we may use transfinite induction to obtain certain properties of $\mathfrak{A}$ -holomorphic functions. For example, the $\mathfrak{A}$ -holomorphic functions are continuous and closed under the ordinary algebraic operations whenever they are defined. Also we have the very important fact that $\mathfrak{A}$ -holomorphic functions satisfy a local maximum modulus

principle. It may be stated as follows:
Let Ω be a compact $\mathfrak{A}$ -convex set in Σ and let U be a relatively open subset of Ω . Also let h be continuous on the closure $\overline{U}$ and $\mathfrak{A}$ -holomorphic on U . Then the maximum modulus of h on U is equal to its maximum modulus on
$(\mathrm{bdry}_{\Omega} U) \cup (\overline{U} \cap \partial_{\mathfrak{A}} \Omega)$. Note that h is only required to be defined on $\overline{U}$. The proof of this principle rests ultimately on the Rossi local maximum modulus principle which holds for $[\Omega, \mathfrak{A}]$. [10, Lemma 2.5].

In the case of $[\mathbb{C}^n, \mathcal{P}]$, every $\mathcal{P}$ -holomorphic function defined on an open set is already $\mathcal{P}$ -holomorphic of class 1 and the collection of all such functions coincides with the ordinary holomorphic functions. An open question for general $\mathfrak{A}$ -holomorphic functions concerns the possible values for the ordinal μ in the decomposition of $\mathfrak{A}_{loc}$. There are examples that require $\mu = 2$ [9], but beyond this nothing seems to be known, although it appears plausible that examples should exist with more-or-less arbitrary values of μ .

Before continuing with the general case, we examine the example $[\mathbb{C}^1, \mathcal{P}]$ in some detail.

§3. Some results for $[\mathbb{C}^{\Lambda}, \mathcal{P}]$.

Consider a commutative Banach algebra $\mathfrak{L}$ (with 1) and let $b \longrightarrow \hat{b}$ denote the Gelfand representation of $\mathfrak{L}$ on its space of maximal ideals $\Phi_{\mathfrak{L}}$. If $\{z_\lambda : \lambda \in \Lambda\}$ is a system of generators for $\mathfrak{L}$, in the sense that polynomials in the generators are dense in $\mathfrak{L}$, then the mapping $\varphi \longrightarrow \{\hat{z}_\lambda(\varphi)\}$ is a homeomorphism of $\Phi_{\mathfrak{L}}$ into $\mathbb{C}^{\Lambda}$. The image $\tilde{\Phi}_{\mathfrak{L}}$ of $\Phi_{\mathfrak{L}}$ in $\mathbb{C}^{\Lambda}$ is $\mathcal{P}$-convex and

$$\tilde{\Phi}_{\mathfrak{L}} = \{\check{\zeta} : \check{\zeta} = \{\zeta_\lambda\} \in \mathbb{C}^{\Lambda}, \ |\zeta_\lambda| \leq \|z_\lambda\|, \ \lambda \in \Lambda\} .$$

Elements of $\hat{\mathfrak{L}}$ go over into functions that are uniform limits on $\tilde{\Phi}_{\mathfrak{L}}$ of polynomials [7, p. 151]. This situation, along with some of the properties of $\mathfrak{L}$ which are obtained through application of results from several complex variables, suggests the possibility of extending some of the fundamental theorems for $[\mathbb{C}^n, \mathcal{P}]$ to the infinite case $[\mathbb{C}^{\Lambda}, \mathcal{P}]$. Such extensions are indeed possible and the resulting theorems are directly applicable (via the above representation) to arbitrary Banach algebras. The key to these extensions is the concept of $\mathcal{P}$-holomorphic function (of class 1) as defined in §2. For convenience, we shall refer here to these functions simply as holomorphic functions in $\mathbb{C}^{\Lambda}$.

The condition that a function be holomorphic on an open set in $\mathbb{C}^{\Lambda}$ is quite restrictive. In fact, such functions are locally finitely determined in the sense that they depend locally on only a finite number of the coordinate variables ζ_λ. This property enables us to reduce arguments to the finite dimensional case. One might also consider functions that are only required to be holomorphic in each of the coordinate variables separately. Although the two conditions coincide for finite dimensions, in the infinite case the latter functions are much more difficult to manage. In any case, the holomorphic functions as defined above appear to be precisely the class of functions required for the results which we wish to obtain here. As might be expected, the proofs involve a reduction to the known results for finite dimensions. Proofs will be found in [11].

The first result is a full extension of the Oka-Weil polynomial approximation theorem [4, p.56; 6, 2.7.2].

Theorem. Let Ω be a compact $\mathcal{P}$-convex set in $\mathbb{C}^{\Lambda}$ and let h be a continuous function on Ω. If h admits a holomorphic extension to a neighborhood of Ω in $\mathbb{C}^{\Lambda}$, then h is a uniform limit on Ω of polynomials.

For a Banach algebra $\mathfrak{B}$ which has the "sup norm"

$$\|b\| = \max \{ |\hat{b}(\varphi)| : \varphi \in \Phi_{\mathfrak{B}} \},$$

the above theorem may be thought of as a convenient formulation of the Arens-Calderon-Šilov Theorem [2, Theorem 3.3]. The latter theorem asserts that if $b_1, \ldots, b_n$ is any finite set of elements of $\mathfrak{B}$ and f is a function holomorphic in the usual sense on a neighborhood of the joint spectrum

$$\mathrm{Sp}(b_1, \ldots, b_n) = \{ (\hat{b}_1(\varphi), \ldots, \hat{b}_n(\varphi)) : \varphi \in \Phi_{\mathfrak{B}} \} \text{ in } \mathbb{C}^n,$$

then there exists an element $b \in \mathfrak{B}$ such that

$$\hat{b}(\varphi) = f(\hat{b}_1(\varphi), \ldots, \hat{b}_n(\varphi)), \quad \varphi \in \Phi_{\mathfrak{B}}.$$

To see how this follows from the above theorem, observe first that we can assume that the elements $b_1, \ldots, b_n$ are contained among the generators $\{ z_\lambda : \lambda \in \Lambda \}$. Thus we are concerned with a set $\{ z_\lambda : \lambda \in \pi \}$, where π is a finite subset of Λ. Note that the natural projection τ_π of $\mathbb{C}^{\Lambda}$ onto $\mathbb{C}^{\pi}$ carries the set $\tilde{\Phi}_{\mathfrak{B}}$ onto the joint spectrum of the elements $\{ z_\lambda : \lambda \in \pi \}$ in $\mathbb{C}^{\pi}$. Moreover, if f is holomorphic in the ordinary sense on a neighborhood of $\tau_\pi \tilde{\Phi}_{\mathfrak{B}}$ in $\mathbb{C}^{\pi}$, then it may be regarded in an obvious way as a function on a neighborhood of $\tilde{\Phi}_{\mathfrak{B}}$ in $\mathbb{C}^{\Lambda}$. As such, it is clearly holomorphic so, by the above theorem, is a uniform limit on $\tilde{\Phi}_{\mathfrak{B}}$ of polynomials. But if $\mathfrak{B}$ has the "sup norm", it follows that every such function corresponds to an element of $\mathfrak{B}$ and the desired result follows.

For application to a Banach algebra with an arbitrary norm, we need a norm version of the above approximation theorem. Observe that, if $P \in \mathcal{P}$ and we denote by b_P the element of $\mathfrak{B}$ obtained by substituting for each variable ζ_λ involved in the polynomial P the corresponding generator z_λ, then $P \longrightarrow b_P$ is a homomorphism of the algebra $\mathcal{P}$ into $\mathfrak{B}$. Therefore, if we define $\|P\| = \|b_P\|$ for each $P \in \mathcal{P}$, the result is a pseudo-norm for the algebra $\mathcal{P}$. In other words,

$0 \leq \|P\| < \infty$, $\|\alpha P\| = |\alpha| \|P\|$, $\|P+Q\| \leq \|P\| + \|Q\|$ and $\|PQ\| \leq \|P\| \|Q\|$ for all $P,Q \in \mathcal{P}$ and $\alpha \in \mathbb{C}$, but $\|P\| = 0$ may not imply $P = 0$. With this motivation, we state a general norm version of the approximation theorem from which the Arens-Calderon-Šilov theorem for an arbitrary Banach algebra is readily deduced.

Assume given an arbitrary pseudo-norm for the algebra $\mathcal{P}$ and let $\{P_\mu : \mu \in M\}$ be any subset of $\mathcal{P}$ which contains the coordinate polynomials Z_λ , $\lambda \in \Lambda$, where $Z_\lambda(\zeta) \equiv \zeta_\lambda$. Define

$$\Omega_M = \{\zeta : \zeta \in \mathbb{C}^\Lambda, |P_\mu(\zeta)| \leq \|P_\mu\|, \mu \in M\}$$

and note that Ω_M is a compact $\mathcal{P}$-convex set in $\mathbb{C}^\Lambda$.

<u>Theorem.</u> <u>If a continuous function on Ω_M admits a holomorphic extension to a neighborhood of Ω_M in $\mathbb{C}^\Lambda$, then it is a uniform limit on Ω_M of polynomials that form a Cauchy sequence with respect to the pseudo-norm.</u>

If Ω is an arbitrary compact $\mathcal{P}$-convex set in $\mathbb{C}^\Lambda$, then $|P|_\Omega$ defines a pseudo-norm in $\mathcal{P}$. Moreover, if we index the algebra $\mathcal{P}$, say $\mathcal{P} = \{P_\mu : \mu \in M\}$, then $\Omega_M = \Omega$. In this case the above theorem reduces to the previous theorem. The proof of the above theorem, which is suggested by Hörmander's [5,§3.2] treatment of the Arens-Calderon-Silov theorem, depends on the following theorem which is of independent interest.

<u>Theorem.</u> <u>Let Δ be a compact polydisc in $\mathbb{C}^\Lambda$ and let Γ be a subset of Δ consisting of the common zeros in Δ of a family of polynomials. If a continuous function on Γ admits a holomorphic extension to a neighborhood of Γ in $\mathbb{C}^\Lambda$, then it also admits a holomorphic extension to a neighborhood of Δ</u> .

Next we consider a Cousin I problem in $\mathbb{C}^\Lambda$. Let Ω be a compact $\mathcal{P}$ -convex set in $\mathbb{C}^\Lambda$ and let $\{U_i : i \in I\}$ be a finite covering of Ω by open sets. A collection $\{h_{ij} : i,j \in I\}$ of functions is called a set of <u>Cousin data</u> for the covering if h_{ij} is holomorphic in $U_i \cap U_j$, $h_{ij} + h_{ji} = 0$, and $h_{ij} + h_{jk} + h_{ki} = 0$ in $U_i \cap U_j \cap U_k$, for all $i,j,k \in I$. The associated Cousin I problem is to obtain functions h_i, holomorphic in U_i, such that $h_i - h_j = h_{ij}$ in $U_i \cap U_j$ for all

$i,j \in I$ [5, p.132]. In order to obtain a solution in the infinite case, we must settle for a bit less.

Theorem. *There exists an open covering* $\{V_i : i \in I\}$ *of* Ω *and functions* $\{h_i : i \in I\}$ *such that* $V_i \subseteq U_i$, h_i *is holomorphic on* V_i, *and* $h_i - h_j = h_{ij}$ *on* $V_i \cap V_j$ *for all* $i,j \in I$.

Following an argument suggested by Stolzenberg [14, (1.8)] for the finite dimensional case, we may use the above solution of Cousin I in to give a direct proof of the local maximum modulus principle for general Banach algebras.

Finally, we denote by ${}_{\Lambda}\mathcal{O}$ the sheaf of germs of holomorphic functions at points of $\mathbb{C}^{\Lambda}$. For an arbitrary subspace X in $\mathbb{C}^{\Lambda}$, let $H^p(X, {}_{\Lambda}\mathcal{O})$ be the pth Cech cohomology group of the space X with coefficients in ${}_{\Lambda}\mathcal{O}$ (restricted to X). Then we have the following theorem which extends to $\mathbb{C}^{\Lambda}$ the well-known result for polynomially convex sets in $\mathbb{C}^n$.

Theorem. *If* Ω *is a compact* $\mathcal{P}$*-convex set in* $\mathbb{C}^{\Lambda}$, *then* $H^p(\Omega, {}_{\Lambda}\mathcal{O}) = 0$ *for* $p \geq 1$.

Arens [1, Theorem 5.1] has obtained a similar result for the space of maximal ideals of a Banach algebra $\mathcal{L}$ in terms of the sheaf of germs of certain functions which he defines as holomorphic in the conjugate space $\mathcal{L}'$ of $\mathcal{L}$ regarded as a Banach space. (Note that $\Phi_{\mathcal{L}}$ is a subset of $\mathcal{L}'$.) It is not difficult to obtain the Arens result as a corollary of the above theorem. In both cases, of course, the proofs depend on the result for $\mathbb{C}^n$.

§4. $\mathfrak{A}$ - analytic varieties.

We return now to the case of an arbitrary natural system $[\Sigma, \mathfrak{A}]$. A continuous function will be called almost $\mathfrak{A}$-holomorphic if it is $\mathfrak{A}$-holomorphic on that portion of its domain of definition where it is non-zero. Almost $\mathfrak{A}$-holomorphic functions satisfy the same local maximum modulus principle that we stated for $\mathfrak{A}$-holomorphic functions. In certain special cases [3], almost $\mathfrak{A}$-holomorphic functions are in fact $\mathfrak{A}$-holomorphic.

Let G be any set in Σ and let Θ be a subset of G. Then Θ is called an $\mathfrak{A}$-analytic subvariety of G if for each $\sigma \in G$ there exists a neighborhood $\mathcal{N}_\sigma$ of σ such that the set $\Theta \cap \mathcal{N}_\sigma$ consists of the common zeros of some family (perhaps infinite) of functions that are defined and almost $\mathfrak{A}$-holomorphic on $G \cap \mathcal{N}_\sigma$. Note that this condition implies that Θ is relatively closed in G. We have the following important result which includes a similar result for finite dimensions [4. VII, A7;14, p. 285.]

Theorem. If Θ is an $\mathfrak{A}$-analytic subvariety of an $\mathfrak{A}$-convex set in Σ, then Θ is also $\mathfrak{A}$-convex.

Since the definition of an $\mathfrak{A}$-analytic subvariety is clearly local and $\mathfrak{A}$-convexity is a global property, the above theorem provides another illustration of an analytic type result for general function algebras. Its proof [10] depends on an extension of a lemma that Glicksberg used to prove a generalization of a theorem of Rado [3, Lemma 2.1]. The Glicksberg lemma depends on the Rossi local maximum modulus principle while the lemma used here depends on the local maximum modulus principle for almost $\mathfrak{A}$-holomorphic functions.

We may also state at this point another analytic type result that generalizes a known result for finite dimensions. It is also an example of local properties implying global properties. Let G be an open set in Σ and let H be a continuous complex-valued function defined on $[0,1] \times G$ such that, for each $t \in [0,1]$, the function h_t, where $h_t(\sigma) \equiv H(t,\sigma)$, is almost $\mathfrak{A}$-holomorphic in G . Also set

$$\Theta_t = \{\sigma : \sigma \in G , \; H(t,\sigma) = 0\} .$$

Then $\{\Theta_t : t \in [0,1]\}$ is called a <u>continuous family of</u> $\mathfrak{A}$ <u>-analytic hypersurfaces</u> in G [14, p. 264]. We say that $\{\Theta_t\}$ <u>intersects</u> a set X if $\Theta_t \cap X \neq \emptyset$ for some t and that it <u>intersects</u> X <u>non-trivially</u> if $\Theta_t \cap X$ is closed in X for each t and the set $\{t : \Theta_t \cap X \neq \emptyset\}$ is a proper, closed, non-empty subset of [0,1].

<u>Theorem.</u> <u>Let</u> K <u>be a compact subset of</u> Σ . <u>Then every continuous family of</u> $\mathfrak{A}$ <u>-analytic hypersurfaces that intersects</u> $\hat{K}$ <u>non-trivially must also intersect</u> K .

This theorem generalizes a result due to Oka [6, pp. 13,14; 14, p. 264] for polynomially convex sets in $\mathbb{C}^n$. Its proof [8; Theorem 4.2] also depends on the local maximum modulus principle along with the facts that in a Banach algebra a limit of regular elements is either regular or a topological divisor of zero and that a topological divisor of zero in a function algebra must have a zero on the Šilov boundary [7, pp. 22, 137].

§5. $\mathfrak{A}$ -holomorphic convexity.

We now consider the generalization of another well-known concept for finite dimensions. Let G be an open set in Σ and denote by $\mathcal{O}_G$ the algebra of all $\mathfrak{A}$-holomorphic functions defined in G . Consider a compact set $K \subset G$ and set

$$\hat{K} = \{\sigma : \sigma \in G , |n(\sigma)| \leq |n|_K , n \in \mathcal{O}_G \} .$$

It is obvious that $\hat{K} \subseteq \hat{K}$, so $\hat{K}$ has a compact closure although it need not be compact itself. In case the set $\hat{K}$ is compact for every compact set $K \subset G$, then G is said to be $\mathfrak{A}$ -holomorphically convex. [10, Def. 3.1]

The notion of holomorphic convexity arises for the finite dimensional case in the study of domains of holomorphy; i.e. domains on which there is defined a holomorphic function which does not admit a holomorphic extension across the boundary of the domain. In fact, for a domain in $\mathbb{C}^n$ to be a domain of holomorphy it is necessary and sufficient that it be holomorphically convex [5, Theorem 2.5.5]. Although the necessity portion of this result is difficult the sufficiency is relatively easy and may be generalized. Thus if G is assumed to satisfy certain countability conditions and is $\mathfrak{A}$ -holomorphically convex, then it is a domain of $\mathfrak{A}$-holomorphy. In fact, there will exist an element of $\mathcal{O}_G$ which is unbounded in every neighborhood of each boundary point of G and so cannot be extended across the boundary [10, Theorem 3.7].

Our main result for $\mathfrak{A}$-holomorphic convexity is given by the following theorem [10, Cor. 3.6].

Theorem. In order for an open set G in Σ to be $\mathfrak{A}$ -holomorphically convex it is necessary and sufficient that $[G, \mathcal{O}_G]$ be natural.

If $[G, \mathcal{O}_G]$ is natural, then for compact $K \subset G$, the set $\hat{K}$ is simply the $\mathcal{O}_G$-convex hull of K and so is automatically compact by the fundamental property of natural systems. This is the sufficiency of the condition. The necessity, which reduces to a known result for finite dimensions [14, (A.21), p. 285], is more difficult. The proof involves a generalization of a technique used by Oka [6, p. 21] and depends on the convexity theorem for $\mathfrak{A}$-analytic subvarieties stated in §4. We sketch the argument here and refer to [10] for details.

Let $\{h_\lambda : \lambda \in \Lambda\}$ denote an indexing of the elements of $\mathcal{O}_G$ and consider the product space $\Sigma \times \mathbb{C}^\Lambda$. Denote by $\mathfrak{A} \times \mathcal{P}$ the Kronecker product of the algebra $\mathfrak{A}$ with the algebra $\mathcal{P}$ of all polynomials on $\mathbb{C}^\Lambda$ (i.e. $\mathfrak{A} \times \mathcal{P}$ consists of all polynomials in a finite number of the variables ζ_λ with coefficients in $\mathfrak{A}$). Then the system $[\Sigma \times \mathbb{C}^\Lambda, \mathfrak{A} \times \mathcal{P}]$ is natural. Next consider the mapping $\sigma \longrightarrow (\sigma, \{h_\lambda(\sigma)\})$ of G into $\Sigma \times \mathbb{C}^\Lambda$. This is the generalized Oka mapping. Denote the image of G in $\Sigma \times \mathbb{C}^\Lambda$ by $\tilde{G}$. Now consider a compact set $K \subset G$ and define

$$\Delta_K = \{\zeta : |\zeta_\lambda| \leq |h_\lambda|_K, \ \lambda \in \Lambda\}.$$

Then Δ_K is a compact polydisc in $\mathbb{C}^\Lambda$. In particular, Δ_K is $\mathcal{P}$-convex. The image of $\hat{K}$ in $\tilde{G}$ is a compact $\mathfrak{A} \times \mathcal{P}$-analytic subvariety of $G \times \Delta_K$. Since G is open and $\hat{K}$ is compact one can show that the image of $\hat{K}$ is an $\mathfrak{A} \times \mathcal{P}$-analytic subvariety of the $\mathfrak{A} \times \mathcal{P}$-convex set $\Sigma \times \Delta_K$ and so, by the convexity theorem, must also be $\mathfrak{A} \times \mathcal{P}$-convex. Since this holds for every compact set $K \subset G$, it follows that $\tilde{G}$ is $\mathfrak{A} \times \mathcal{P}$-convex. Finally, the $\mathfrak{A} \times \mathcal{P}$-convexity of $\tilde{G}$ may be used to extablish the naturality of $[G, \mathcal{O}_G]$.

Bibliography

1. R. Arens. The problem of locally A-functions in a commutative Banach algebra A. Trans. Amer. Math. Soc. 104 (1962), 24-36.

2. R. Arens and A. P. Calderon. Analytic Functions of several Banach algebra elements. Ann. of Math. 62 (1955), 204-216.

3. I. Glicksberg. Maximal algebras and a theorem of Rado. Pacific J. Math. 14(1964), 919-941.

4. R. C. Gunning and H. Rossi. Analytic Functions of Several Complex Variables. Prentice-Hall, Englewood Cliffs, N. J., 1965.

5. L. Hörmander. An Introduction to Complex Analysis in Several Variables. D. Van Nostrand Co., Princeton, N. J., 1966.

6. K. Oka. Sur les functions analytiques de plusieures variables Iwanami Shoten, Tokyo, 1961.

7. C. E. Rickart. General Theory of Banach Algebras. D. Van Nostrand Co., Princeton, N. J., 1960.

8. ____________. Analytic phenomena in general function algebras. Pacific J. Math. 18 (1966), 361-377. Proc. Tulane Symposium on Function Algebras (pp. 61-64) (F. Birtel, Editor), Scott, Foresman and Co., Chicago, Ill., 1966.

9. ____________. The maximal ideal space of functions locally approximable in a function algebra. Proc. Amer. Math. Soc. 17 (1966), 1320-1326.

10. ____________. Holomorphic convexity in general function algebras. Canadian J. Math. (To appear).

11. ____________. Analytic functions of an infinite number of variables. (Submitted for publication.)

12. H. Rossi. The local maximum modulus principle. Ann. of Math. 72 (1960), 1-11.

13. G. Stolzenberg. A hull with no analytic structure. Journal of Math. and Mech. 12 (1963), 103-112.

14. ____________. Polynomially and rationally convex sets. Acta Math. 109 (1963), 259-289.

THE CAUCHY PROBLEM FOR DIFFERENTIAL EQUATIONS WITH CONSTANT COEFFICIENTS

Lars Hörmander

1. Introduction. The aim of this lecture is to survey what is known concerning uniqueness and existence of solutions of differential equations with constant coefficients with Cauchy data given on a hyperplane. The main emphasis will be placed on recent work on the case where the hyperplane is characteristic, but for contrast we shall also discuss the main facts about the non-characteristic case.

A differential operator in $\mathbb{R}^n$ with constant coefficients can be written in the form $P(D)$ where P is a polynomial in n variables with complex coefficients and $D = (-i\partial/\partial x_1, \ldots, -i\partial/\partial x_n)$. Let m be the degree of $P(\xi)$ with respect to ξ and let μ be the degree with respect to ξ_n for generic $\xi' = (\xi_1, \ldots, \xi_{n-1})$. The Cauchy problem then consists, roughly speaking, in finding for given functions f in $\mathbb{R}^n_+ = \{x;\ x \in \mathbb{R}^n,\ x_n \geq 0\}$ and ϕ_j in $\mathbb{R}^{n-1} = \partial\mathbb{R}^n_+$, $j < \mu$, a function u in $\mathbb{R}^n_+$ such that

$$P(D)u = f \quad \text{in } \mathbb{R}^n_+ \quad \text{and} \quad D_n{}^j u = \phi_j \quad \text{in } \partial\mathbb{R}^n_+ \quad \text{for } j = 0, \ldots, \mu-1.$$

If $f \in C^\infty(\mathbb{R}^n_+)$ and $\phi_j \in C^\infty(\mathbb{R}^{n-1})$, $j < \mu$, one can successively find functions $\phi_j \in C^\infty(\mathbb{R}^{n-1})$ for $j \geq \mu$ such that if $v \in C^\infty(\mathbb{R}^n_+)$ and $D_n{}^j v = \phi_j$ on $\partial\mathbb{R}^n_+$, $j \geq 0$, it follows that the difference $P(D)v - f$ vanishes of infinite order when $x_n = 0$. If u is a solution of the Cauchy problem and we put $u_0 = u - v$, $f_0 = f - P(D)v$ in R^n_+, extending u_0 and f_0 to be 0 outside R^n_+, it follows that $f_0 \in C^\infty(\mathbb{R}^n)$ and that $P(D)u_0 = f_0$ (in the sense of distribution theory at least). Conversely, if $u_0 \in C^{\mu-1}(\mathbb{R}^n)$ satisfies the equation $P(D)u_0 = f_0$ and supp $u_0 \subset \mathbb{R}^n_+$, it follows that $u = v + u_0$ is a solution of the Cauchy problem. Thus we are led to the following convenient statement of the Cauchy problem, which has the advantage that it still makes sense when no strong smoothness assumptions are fulfilled:

Given a function (distribution) f *in* $\mathbb{R}^n$ *with support in* R^n_+, *find a function (distribution)* u *in* $\mathbb{R}^n$ *with support in* R^n_+ *such that* $P(D)u = f$. In this setting the uniqueness problem is reduced to deciding if there is a solution of the homogeneous equation $P(D)u = 0$ with support in R^n_+.

2. *Uniqueness theorems*. The plane $x_n = 0$ is called *non-characteristic* if $m = \mu$ with the notations of the introduction, that is, the coefficient of D_n^m in $P(D)$ is not equal to 0. In that case it was proved already around 1900 by E. Holmgren that there is *no non-trivial solution of the equation* $P(D)u = 0$ *with support in* R^n_+; his theorem also implies local uniqueness if u is only defined in part of R^n. For smooth u the proof is based on integrations by parts in the integral

$$0 = \int_\Omega (P(D)u)\, v\, dx$$

where Ω is an open set bounded by a part of the plane $x_n = 0$ and a part of a sphere close to that plane. If one can choose v so that $P(-D)v = 0$ and v has suitable Cauchy data on the spherical part of the boundary it is possible to conclude that u is equal to 0 there. The crucial point in the proof is that the Cauchy-Kovalevsky theorem guarantees the existence of solutions of the equation $P(-D)v = 0$ with suitably prescribed analytic Cauchy data. For the details of the proof we refer to [3, section 5.3]; it is valid for operators with analytic coefficients. For operators with constant coefficients a proof based on Fourier transforms and analytic function theory will be given in a book by Ehrenpreis to be published soon.

The following result by Hörmander [4] shows that in the characteristic case the Cauchy data give no solid information about the solution: <u>For every P having no factor for which $\partial\mathbb{R}^n_+$ is non-characteristic</u>, <u>the closure in</u> $C^\infty(H)$, $H = \{x;\ x_n > 0\}$, <u>of restrictions to</u> H <u>of solutions of</u> $P(D)u = 0$ <u>in</u> $C^\infty(\mathbb{R}^n)$ <u>with</u> supp $u \subset R^n_+ = \overline{H}$ <u>consists of all solutions of the equation</u> $P(D)u = 0$ in $C^\infty(H)$.

Even in the characteristic case we may have uniqueness for the Cauchy problem if growth conditions are imposed at infinity. As an example consider a solution of the heat equation in R^{n+1}

$$\Delta u - \partial u/\partial t = 0$$

when $t \geq 0$. If $|u(x,t)| \leq C\, e^{A|x|^2}$ it is easy to justify the formula

$$u(x,t) = (4\pi t)^{-n/2}\int e^{-|x-y|^2/4t}\, u(y,0)\, dy,\ 0 < t < 1/4A,$$

and so one can in particular conclude that a solution u of the Cauchy problem with vanishing Cauchy data for $t = 0$ must vanish identically if $|u(x,t)| \leq Ce^{A|x|^2}$. This elementary fact has been extended in two ways. For the heat equation Täcklind [7] proved that if ϕ <u>is a positive increasing function on</u> $(0,\infty)$ <u>then every solution of the heat equation for</u> $0 \leq t \leq T$ <u>which vanishes when</u> $t = 0$ <u>and satisfies the estimate</u>

$|u(x,t)| \leq Ce^{A|x|\phi(|x|)}$ <u>must vanish identically if and only if</u>

$$\int_1^\infty dt/\phi(t) = \infty \quad .$$

On the other hand, Gelfand and Shilov [2] have extended the elementary uniqueness theorem for the heat equation to more general equations. They proved for example that if

$$|\zeta_n| \leq C(1+|\zeta'|)^p \quad \text{when} \quad \zeta = (\zeta',\zeta_n) \in \mathbb{C}^n \text{ and } P(\zeta) = 0 ,$$

where $p > 1$, and we set $1/p + 1/p' = 1$, then there is <u>uniqueness for the Cauchy problem among functions such that</u> $|u(x)| \leq C \exp A|x'|^{p'}$. The proof is parallel to that of the Holmgren uniqueness theorem; one integrates by parts in

$$\int_\Omega (P(D)u)\ v\ dx$$

where Ω is a slab $0 < x_n < t$ and v is chosen as a solution of $P(-D)v = 0$ in Ω with Cauchy data for $x_n = t$ belonging to a suitable class of test functions for which the Cauchy problem can be solved by means of the Fourier transformation and the integration by parts can be carried out without difficulties at infinity. For details we refer to [1] and [2]. For general operators there do not seem to exist in the literature any uniqueness theorems which are as precise as those proved by Täcklind for the heat equation, but Ehrenpreis has announced that such results will appear in his book.

3. <u>Hyperbolic operators</u>. If the Cauchy problem can be solved there exists in particular a solution of the equation $P(D)E = \delta$, where δ is

the Dirac measure at 0 , such that supp $E \subset R^n_+$. We recall that E is then called a fundamental solution. If the plane $x_n = 0$ <u>is non-characteristic</u> it follows from Holmgren's uniqueness theorem not only that there can only be one such E but also that the support of E must belong to a closed cone contained in the interior of R^n_+ apart from the vertex at 0. When such a fundamental solution exists one can solve the equation $P(D)u = f$ for any f with support in R^n_+ by forming the convolution $u = E * f$, which is well defined and has support in R^n_+ . Thus the Cauchy problem can always be solved then, and we say that $P(D)$ is <u>hyperbolic</u> with respect to the plane $x_n = 0$.

A necessary condition for the existence of E can be obtained as follows. Let u be the product of E and a C_0^∞ function which is 1 in a neighborhood of 0. Then $P(D)u = \delta - g$ where the support of g is a compact subset of the interior of R^n_+ . It follows that

$$v(0) = g(v) \quad \text{if} \quad v \in C^\infty \quad \text{and} \quad P(-D)v = 0 . \tag{3.1}$$

In particular, if $P(\zeta) = 0$ we can set $v(x) = \exp -i\langle x,\zeta\rangle$ and obtain for some $\delta > 0$

$$1 = |\hat{g}(\zeta)| \leq C(1+|\zeta|)^N \exp (A|\operatorname{Im} \zeta'|+\delta \operatorname{Im} \zeta_n) \quad \text{if} \quad P(\zeta) = 0 \quad \text{and} \quad \operatorname{Im} \zeta_n < 0 . \tag{3.2}$$

Here we have used the standard estimates for the Fourier transform of a distribution with compact support. If ζ' is real we conclude that

$$\operatorname{Im} \zeta_n \geq -b \log (1+|\zeta|) - c$$

for some constants b and c . Using the algebraic nature of ζ_n one can deduce that for some constant C

(3.3) $\operatorname{Im} \zeta_n \geq -C$ if $P(\zeta) = 0$ and ζ' is real.

Since the sum of the zeros of $P(\zeta',\zeta_n)$ is a linear function of ζ', it follows from (3.3) that

(3.4) $|\operatorname{Im} \zeta_n| \leq C'$ if $P(\zeta) = 0$ and ζ' is real.

Together with the hypothesis that the plane $x_n = 0$ is non-characteristic, this is _Garding's definition of hyperbolicity_. (See e.g. [3, sections 5.4 - 5.6].) We shall sketch a proof of the sufficiency in the next section.

When $P(D)$ is not hyperbolic it is quite exceptional that the Cauchy problem can be solved. This is shown very clearly by the following theorem of F. John (see [3, section 5.7]): _Suppose that the plane $x_n = 0$ is non-characteristic with respect to $P(D)$ and that no factor of $P(D)$ has a principal part which is hyperbolic with respect to that plane. Then there is no solution of $P(D)u = 0$ in R^n_+ which has Cauchy data of compact support not identically 0._

Note that if an operator is hyperbolic it follows that the principal part is hyperbolic but not conversely. (The principal part is the sum of the terms of order m in $P(D)$.) However, assuming only that the principal part is hyperbolic one can easily show that the Cauchy problem can be solved for arbitrary data in suitable Gevrey classes, thus for many data with compact support. See [2, Chapter III, section 5.2], [3, section 5.7] or [6].

4. _Correctness in the sense of Petrowsky._ We shall now drop the assumption that the plane $x_n = 0$ is non-characteristic which was made in

section 3, but we still require that the coefficient of the highest power D_n^μ of D_n in $P(D)$ is constant. (Thus we assume partial hypoellipticity with respect to the plane $x_n = 0$; see [3, section 4.2].) The operator is then said to be correct in the sense of Petrowsky if (3.3) is valid. If $\hat{E}(\xi', x_n)$ is the fundamental solution of the ordinary differential operator $P(\xi', D_n)$ which vanishes for $x_n < 0$, it follows easily from (3.3) that for suitable constants C' and N

$$|\hat{E}(\xi', x_n)| \leq C' e^{Cx_n} (1+x_n)^{\mu-1}(1+|\xi'|)^N, \; x_n \geq 0 .$$

Hence the inverse Fourier transform of $\hat{E}$ with respect to ξ' is well defined in the sense of Schwartz and gives a fundamental solution E with support in R_+^n. More generally we conclude that the Cauchy problem is always uniquely solvable in the appropriate spaces of tempered distributions when (3.3) is valid.

The preceding result suggests the following questions:

a) Is (3.3) necessary for the operator $P(D)$ to have a fundamental solution with support in R_+^n ?

b) Does (3.3) imply that the equation $P(D)u = f$ has a solution with support in R_+^n if f has its support in R_+^n but is allowed to grow rapidly at infinity?

The answer to question a) is negative. In fact, if $P(D)E = \delta$ it follows that $P(D+\vartheta)(e^{-i\langle x,\vartheta\rangle}E) = \delta$ for any $\vartheta \in \mathbb{C}^n$, so the class of operators having a fundamental solution with support in R_+^n is invariant under translations. However, $P(D)$ may satisfy (3.3) while $P(D+\vartheta)$ fails to do so. An example is given by the Schrodinger equation corresponding to $P(\xi) = \xi_2-\xi_1^2$ if we take $\vartheta = (i,0)$. On the other hand, Malgrange pointed out about ten year ago that methods used in his thesis can be modified to show that the equation $P(D)u = f$ has a solution u of finite order with support in R_+^n for every f of finite order with support in R_+^n provided that there exists a fundamental solution of finite order with support in R_+^n. The proof is included in [3, section 5.8] in a slightly special case. In particular, the answer to question b) is therefore affirmative.

5. Evolution operators. We shall say that $P(D)$ *is an evolution operator* with respect to R_+^n if there exists a fundamental solution with support in R_+^n. (Note that we no longer assume that $P(D)$ is partially hypoelliptic with respect to the hyperplane $x_n = 0$.) Thus we have

Hyperbolic operators $\subset$ operators correct in the sense of Petrowsky $\subset$ evolution operators.

The essential gap in the last inclusion is caused by the lack of translation invariance of the Petrowsky condition pointed out above. Indeed, in [5] we have recently characterized evolution operators as follows:

Theorem 1. The following conditions on $P(D)$ *are equivalent*:

(i) $P(D)$ *is an evolution operator with respect to* R_+^n.

(ii) *There exist constants* A_1 *and* A_2 *with* $A_1 > 0$ *such that for every solution* $\zeta_n(\zeta')$ *of the equation* $P(\zeta',\zeta_n) = 0$ *which is analytic and single valued in a ball* B *with real center and radius* A_1 *we have*

$$\sup_B \operatorname{Im} \zeta_n(\zeta') \geq A_2 .$$

(iii) *The equation* $P(D)u = f$ *has a solution* u *of finite order with support in* R_+^n *for every* f *of finite order with support in* R_+^n.

As mentioned above the equivalence of (i) and (iii) is essentially due to Malgrange. The results in [5] are much more precise than the statement of Theorem 1 here so they permit one to prove the stability of the class of evolution operators in the following sense.

Theorem 2. _If_ $P(D)$ _is an evolution operator it follows that_ $Q(D)$ _is an evolution operator if_ Q _is in the component containing_ $P(D)$ _of the space of operators which are equally strong as_ $P(D)$.

Note that the operators weaker than $P(D)$ form a finite dimensional vector space W where the set of operators which are equally strong as $P(D)$ is open. (See [3, section 3.3] for the concepts involved here.)

The proof of Theorem 1 is fairly technical so we shall only sketch it very briefly. As in section 3 the implication (i) $\Rightarrow$ (ii) follows from (3.1). However, g may now have points with $x_n = 0$ in its support although the origin does not belong to the support. All exponentials may therefore satisfy (3.1), so we must apply (3.1) to more complicated functions obtained by a smooth averaging of exponential solutions $\exp(i\langle x', \xi'\rangle + \zeta_n(\xi')\rangle$ where ζ_n is a function with the properties listed in condition (ii). The fact that Fourier transforms of smooth functions are small at infinity then tends to decrease the contribution of $g(v)$ coming from points in the support of g where $x_n = 0$. The necessity of averaging exponential solutions explains why condition (ii) is not a condition on any individual zero of P but only restricts the location of large pieces of the surface $P(\zeta) = 0$.

The implication (iii) $\Rightarrow$ (i) is trivial so it remains to discuss the proof that (ii) $\Rightarrow$ (iii). Arguing by duality one has to prove that if $\phi \in C_0^\infty(K)$ where K is a compact subset of R^n, then the restriction of ϕ to R_-^n can be estimated in terms of the restriction of $\psi = P(D)\phi$ to R_-^n. After taking Fourier transforms in the x' variables we have the ordinary differential equation

$$(5.1) \qquad P(\zeta', D_n)\, \hat{\phi}(\zeta', x_n) = \hat{\psi}(\zeta', x_n) .$$

For each ζ' where a zero of $P(\zeta',\tau)$ satisfies $\operatorname{Im} \tau \geq A_2$ we can remove a factor $(D_n - \tau)$ by an integration while preserving all reasonable estimates of the right hand side when $x_n < 0$. The removal of arbitrary factors is then based on the following observations:

1° If $\zeta_n(\zeta')$ is the function in condition (ii) and ζ_0' is a point where $\operatorname{Im} \zeta_n(\zeta_0') \geq A_2$, then we have $\operatorname{Im} \zeta_n(\zeta') \geq A_2$ for some point on the boundary of any complex disc with center at ζ_0' contained in the domain of ζ_n. Thus (ii) guarantees that there are many points ζ' where $\operatorname{Im} \zeta_n(\zeta') \geq A_2$.

2° Let f be an analytic function in the unit disc in $\mathbb{C}$ with maximum modulus $\leq M$ and minimum modulus $\leq m$, that is,

$$|f(z)| \leq M \text{ when } |z| < 1, \quad \inf_{\theta} |f(re^{i\theta})| \leq m \text{ when } 0 \leq r < 1.$$

Then it follows that

$$|f(z)| \leq M^{1-\delta} m^{\delta}$$

where $\delta = \delta(|z|)$ is a positive continuous function when $0 \leq |z| < 1$, which is independent of f. In fact, the best value for δ is known explicitly from the classical work of E. Schmidt, R. Nevanlinna and A. Beurling.

3° If $u \in C_0^\infty(R^{n-1})$ has support in the set where $|x'| \leq A$ it follows from the Phragmen-Lindelöf theorem that

$$|\hat{u}(\zeta')|\, e^{-A|\operatorname{Im}\zeta'|} \leq \sup_{\xi'} |\hat{u}(\xi')|, \quad \zeta' \in \mathbb{C}^{n-1}, \ \xi' \in \mathbb{R}^{n-1} .$$

Let $\Psi(\zeta', x_n)$ denote the μ-valued function obtained by removing a linear factor from (5.1). To estimate Ψ one notices first that in view of 1° it is possible to get bounds for the minimum modulus in a branch of Ψ centered at a point where the zero removed lies in the half plane $\mathrm{Im}\ \zeta_n \geq A_2$. One can then use 2° to get an estimate for maximum modulus. The fact that application of 2° requires one to go to a smaller disc does not matter since by 3° norms obtained by letting ζ' vary over different strips around R^{n-1} are largely independent of the width of the strip. One is therefore led to estimates of the form

$$||\Psi|| \leq C||\Psi||^{1-\delta}||\psi||^{\delta}$$

for some $\delta > 0$, hence $||\Psi|| \leq C'||\psi||$. By repeating the argument one gets finally $||\phi|| \leq C||\psi||$ if for example

$$||\phi|| = \sup_{x_n<0} \sup_{\xi'} |\hat{\phi}(\xi', x_n)|$$

and $||\psi||$ is defined similarly. Actually much more is proved in one stroke in [5] but it would take too long to go into the full technical details here. In particular, one has to take care to avoid coming close to zeros of the coefficient of ζ_n^{μ} in $P(\zeta)$ or to zeros of the discriminant of an irreducible factor of P. By standard duality arguments the inequality $||\phi|| \leq C||\psi||$ implies the existence statement (iii) in Theorem 1.

References

1. A. Friedman, Generalized functions and partial differential equations. Prentice Hall, New York, 1962.

2. I. M. Gelfand and G. E. Shilov, Generalized functions, Vol. 3: Theory of differential equations. Moscow 1958 (Russian), Academic Press, New York, 1967.

3. L. Hörmander, Linear partial differential operators. Springer Verlag 1963.

4. - , Null solutions of partial differential equations. Arch. Rational Mech. Anal. 4 , 255-261 (1960).

5. - , On the characteristic Cauchy problem. To appear in Ann. of Math.

6. J. Leray and Y. Ohya, Systemes lineaires, hyperboliques non stricts. Seminaire sur les equations aux derivees partielles, College de France 1964, 20 - 71.

7. S. Täcklind, Sur les classes quasianalytiques des solutions des equations aux derivees partielles du type parabolique. Nova Acta Soc. Sci. Upsaliensis (4) 10, 1-57 (1936).

Local Cauchy Problem for Linear Partial Differential Equations with Analytic Coefficients

F. Treves

1. Introduction

Let $\mathcal{O}$ be an open subset of the Euclidean space R^{n+1}, P a linear partial differential operator of order m, with coefficients defined and analytic in $\mathcal{O}$, S, a piece of analytic hypersurface in $\mathcal{O}$, x^o a point in S. We denote by $\partial/\partial\nu$ the differentiation in the normal direction to S at a point of S (we may assume that S is diffeomorphic to a hyperplane, and oriented). The local Cauchy problem relative to P and to S about the point x^o consists in determining if for a suitable open neighborhood U of x^o in $\mathcal{O}$, the equations

(1) $\quad Pu = f$ in U ;

(2) $\quad (\partial/\partial\nu)^k u = \phi_k$ on $S \cap U$ for $k = 0, \dots, m-1$,

have a solution; additional questions might be as to whether the solution is unique and whether it depends continuously on the data, f , ϕ_k $(0 \leq k < m)$. If this is so, one says, after Hadamard, that the problem is "well posed." But since we have not stated what the data are, our definition must remain rather vague. In what we are going to say, the hypersurface S will always be <u>noncharacteristic</u> with respect to P at the point x^o : this means that if g is an analytic function defining S near x^o (which implicitly assumes that grad $g \neq 0$ near x^o), then the polynomial in the indeterminate τ ,

$$\{\exp(-\tau g(x))\, P(\exp(\tau g(x))\}_{x=x^o}$$

is effectively of degree m . Under this assumption, when the data f and ϕ_k are analytic, the local Cauchy problem has always a unique solution: this is the classical Cauchy-Kovalevska theorem. It hardly needs recalling that the situation changes radically when the data stop being analytic. The basic example is provided by the Cauchy-Riemann equation $(\partial/\partial\bar{z})u = 0$: its solutions are holomorphic functions of $z = x + iy$, and if we wish to prescribe initial data, say on a piece of straight line in the (x,y)-plane, they better be analytic. There is certainly not going to be any solution if the data are $\mathcal{C}^\infty$ functions with compact support. Let us underline that this is true, even if we are willing to accept distributions solutions: for the distributions solutions of the homogeneous equation $(\partial/\partial\bar{z})u = 0$ are identical with its smooth solutions.

As a matter of fact, we run into serious difficulties even if we drop the "initial conditions," (2), as implied by the "nonexistence" results which have followed the discovery by Lewy that the operator

$$L = (\partial/\partial x_1) + i(\partial/\partial x_2) + i(x_1 + ix_2)(\partial/\partial x_3)$$

was not "solvable" in any open subset Ω of R^n , meaning by this that given arbitrarily a function ϕ in $\mathcal{C}_c^\infty(\Omega)$, there will not be, in general, a distribution u in Ω satisfying $Lu = \phi$ (in Ω). These nonexistence results (see Lewy [1], Hörmander [1], Ch. VI, Nirenberg-Treves [1]) have raised the question as to the desirability of enlarging the "inventory" of possible solutions, from the distributions to perhaps more "generalized" objects. In relation with this, a recent negative result, due to P. Schapira, should be mentioned: <u>given any open neighborhood</u> Ω <u>of</u> 0 <u>in</u> R^2 , <u>there</u>

is at least one $\mathcal{C}^\infty$ function in R^2 , f , such that the equation $(\partial/\partial x_1 + ix_1 \, \partial/\partial x_2)u = f$ has no solution in the space of Sato hyperfunctions (Schapira [1]). Sato hyperfunctions "are" boundary values, in the real space, of holomorphic functions (see Sato [1]).

In the forthcoming, we are going to outline the proof of a result that points in the other direction. This will imply that if we choose appropriately the neighborhood U , then, for any f in $\mathcal{C}_c^\infty(U)$, any Cauchy data ϕ_k in $\mathcal{C}_c^\infty(U \cap S)$ $(0 \leq k < m)$, the problem (1)-(2) has a unique solution; this solution, in general, will not be a distribution: it will be something more general, what we call an ultradistribution; although these objects lack some of the most important properties of distributions, in particular all the properties related with localization (such as support), they retain some important ones, and important operations (such as Fourier transformation, convolution and, for some of them, multiplication) can be performed on them. We shall describe them in some detail.

The existence and uniqueness of the solutions to (1)-(2) is valid for data more general than $\mathcal{C}^\infty$ with compact support; the data can be taken in the spaces of ultradistributions which we are going to describe (these can be made to contain all distributions with compact support, and more generally, distributions in U or in $S \cap U$ which decay sufficiently rapidly at the boundary). The solutions will then belong to sets of ultradistributions whose "degree of regularity" is strictly lesser than (although arbitrarily close to) the one of the solutions: this is in agreement with the fact that one should not get distribution-solutions for arbitrary distributions (or $\mathcal{C}_c^\infty$ functions) as data.

A few words are in order about the uniqueness of the solutions to (1)-(2) ; as it is stated in the main result of this lecture, it implies at once the classical Holmgren's theorem, and its distribution version implies the modern version of Holmgren's theorem (Hörmander [1], Th. 5.3.1). The counter-examples to uniqueness in the Cauchy problem, due to Plis, Cohen et alt., show that there cannot be any extension of our main result to the case of nonanalytic coefficients, unless we are willing to restrict the type of the equation (for hyperbolic equations with $\mathcal{C}^\infty$ coefficients, the problem (1)-(2) is of course well posed in the space $\mathcal{C}^\infty$).

2. The abstract differential equation (*)

The method consists of three rather distinct parts. The first one centers around an "abstract" ordinary linear differential equation, and the related " Cauchy problem":

(3) $$\frac{du}{dt} = A(t)u + f ,$$

(4) $$u/_{t=0} = u_o .$$

There $A(t)$ stands for a matrix with entrico partial differential operators and pseudodifferential operators in the x-variables, with coefficients depending (analytically) on t ; it will act on certain Banach spaces, and as usual in such situations, $A(t)$ will be an unbounded linear operator in these Banach spaces. In most analyses of this type, the Banach spaces are L^p spaces or spaces of Hölder continuous functions, or Sobolev spaces. As an example suppose that we are dealing with the Sobolev spaces H^s , and that $A(t)$ stands for the Laplace operator in the x-variables. Then $A(t)$ will be unbounded in every H^s but will define a bounded linear operator $H^s \to H^{s-2}$. Likewise, we incorporate the Banach spaces on which $A(t)$ is going to act into a one-parameter family of Banach spaces. We denote these by X_s; for simplicity we assume $0 \leq s \leq 1$ and (just as in the case of the H^s spaces)

(5) *If* $0 \leq s' \leq s \leq 1$, $X_s \subset X_{s'}$ *and the injection* $X_s \to X_{s'}$ *has norm* ≤ 1

But the crucial difference between the Sobolev spaces situation and ours lies in the fact that $A(t)$ will define a bounded linear operator from X_s into $X_{s'}$ *whatever* $s' < s$. In the nontrivial cases, the norm of this operator will tend to $+\infty$ as $s' < s$ tends to s; but this will not be allowed to take place in a completely free manner; the growth of the norm of $A(t) : X_s \to X_{s'}$ will have to be of the order of magnitude of $(s - s')^{-1}$. More precisely:

(6) *For all* $s, s', s' < s$, $A(t)$ *is an analytic function of* t, $|t| < \rho$ *valued in the Banach space of bounded linear operators* $X_s \to X_{s'}$ *with norm* $\leq C(s - s')^{-1}$, *where* $C > 0$ *is a constant independent of* s, s' *and* t.

In (6) the word "analytic" can be understood in the complex sense, and we may look on t as a complex variable: localization of the problem with respect to t allows analytic continuation into the complex plane.

Condition (6) is not as artificial as it might seem; the following examples help to understand it better.

Example 1. Let E be a finite dimensional normed space, r_o, r_1 two numbers such that $0 < r_1 < r_o$, and D_s the closed polydisk in C^n defined by the conditions $|z_j| \leq sr_o + (1 - s)r_1$ $(j = 1,\ldots, n)$. We take for X_s the Banach space of continuous mappings $D_s \to E$ which are holomorphic in the interior of D_s ; the norm in X_s is the maximum of the norm in E of the values of the function. Property (5) clearly holds. Consider $(n + 1)$ linear endomorphisms of E , A_j $(0 \leq j \leq n)$, and take

$$A(t) = A_0 + \sum_{j=1}^{n} A_j \, \partial/\partial z_j \, .$$

It follows at once from Cauchy's formula that (6) holds.

Example 2. Let $g(\xi)$ be a continuous positive function in R^n and denote by $L^p(g)$ $(1 \leq p < +\infty)$ the space of (classes of) functions f in R^n such that

$$e^g f \in L^p .$$

Let s be any real number, and take as space X_s the space $L^p(sg)$ (the fact that s varies over the whole real line, rather than in the closed unit interval, is of no importance). Property (5) holds trivially. Let $\phi(\xi,t)$ a holomorphic function of t , $|t| < \rho$, valued in the space of continuous bounded functions of ξ in R^n , and suppose that, for some constant $M > 0$ and all t, $|t| < \rho$, and $\xi \in R^n$, $|\phi(\xi,t)| \leq M\, g(\xi)$. If then $A(t)$ = multiplication by $\phi(\xi,t)$, Property (6) is satisfied. Indeed, if $s' < s$, and $u \in X_s$,

$$\|\phi(.,t)u\|_{X_{s'}} = \|e^{s'g}\phi(.,t)u\|_{L^p} \leq M \, \| g \, e^{-(s-s')g} \, e^{sg} \, u\|_{L^p}$$

$$\leq Me^{-1} \, \|e^{sg} \, u\|_{L^p} \quad \text{since } a\, e^{-a} \leq e^{-1} \text{ for all } a \geq 0.$$

The holomorphy of $A(t)$ is an easy consequence of Cauchy's formulae. The relevance of Example 2 will be clearer when we start making use of Fourier transforms.

At any rate, we may now state the main result concerning the problem (1)-(2) under our assumptions:

Theorem 1. (*) *Suppose that* (5) *and* (6) *hold. Let* $f(t)$ *be a holomorphic function of* t , $|t| < \rho$, *valued in* X_1 *and let* u_o *belong to* X_1 . *Then there is a unique solution* $u(t)$ *of* (1)-(2) *which for some* $\delta > 0$ *and some* s , $0 \leq s < 1$, *is a holomorphic function of* t , $|t| < \delta$, *valued in* X_s .

Then, for every s , $0 \leq s < 1$, $u(t)$ *is a holomorphic function of* t , $|t| < \delta(s) = \alpha(1 - s)$, *valued in* X_s , α *being a constant*, $0 < \alpha < \rho$, *independent of* s .

There is a "real" version of Th. 1: we assume that t is a real variable, varying in the interval $|t| < \rho$, and we replace the word "analytic" in (6) by "continuous," and we assume that $f(t) : \{t ; |t| < \rho\} \to X_s$ is also continuous (and not any more holomorphic). Then the solution $u(t)$ is continuously differentiable. We should also mention that there is a *nonlinear* version of Th. 1 : we substitute for (6) its evident nonlinear generalization (see Treves [2]).

Outline of the proof of Th. 1 : we shall content ourselves with outlining the proof of the existence of the solution; the remaining is routine. One solves the problem

(7) $$v'(t) = A(t)v(t) + g(t) , \quad |t| < \delta , \qquad v(0) = 0 ,$$

where $g(t) = f(t) + A(t)u_o$. Setting afterwards $u = v + u_o$ yields a solution of (1)-(2) (observe that $g(t)$ is a holomorphic function of t, $|t| < \rho$, valued in X_s for all $s < 1$, but not necessarily valued in X_1). Then one defines inductively a sequence of functions $v_k(t)$

as follows:

$$v_0(t) = \int_0^t g(z)dz \ , \qquad v_{k+1}(t) = \int_0^t A(z)v_k(z)\,dz \ ,$$

where the integration is performed over the straight line segment in the complex plane joining 0 to t. If we show that the series $\sum_{k=0}^{+\infty} v_k(t)$ converges uniformly on every compact subset of a disk $|t| < \delta$, its sum $v(t)$ will be clearly a solution of (7). In order to see this, we prove, by induction on $k = 0, 1, \ldots,$ an estimate

(8) $$\|v_k(t)\|_{X_s} \leq B\,(1 - s)^{-k-1}(Ce|t|)^{k+1} \ , \quad |t| < \delta \ , \ k = 0, 1, \ldots,$$

where C is the constant in (6) and B a constant now to be determined. Requiring that δ be $< \eta$, a fixed number such that $0 < \eta < \rho$, we observe that

$$\sup_{|t|<\delta} \|g(t)\|_{X_s} \leq \sup_{|t|\leq\eta} \|f(t)\|_{X_1} + C\,(1 - s)^{-1}\|u_o\|_{X_1} \ .$$

We see thus that we may find B large enough so as to satisfy (8) when $k = 0$. Let now ε be any number such that $0 < \varepsilon < 1 - s$. We have, in view of (6) and of the induction hypothesis,

$$\begin{aligned}\|v_{k+1}(t)\|_{X_s} &\leq \int_0^1 \|A(tr)v_k(tr)\|_{X_s} |t|dr \leq \frac{C}{\varepsilon}|t| \int_0^1 \|v_k(tr)\|_{X_{s+\varepsilon}} dr \\ &\leq B\,C|t|(Ce|t|)^{k+1}\,\varepsilon^{-1}\,(1 - s - \varepsilon)^{-k-1} \int_0^1 r^{k+1}\,dr \ .\end{aligned}$$

Choosing $\varepsilon = (1 - s)/(k + 2)$ yields

$$\|v_{k+1}(t)\|_{X_s} \leq B\,(1 - s)^{-k-2}\,C|t|\,(Ce|t|)^{k+1}\,(1 + \frac{1}{1 + k})^{k+1} \ ,$$

whence (8) for $(k + 1)$ instead of k.

One can recognize the technique used in the proof of the Cauchy-Kovalevska theorem in Hörmander [1], p. 118. The advantage of a theorem such as Th. 1 resides of course in its generality, and in the flexibility of its applications. We may apply it not only to spaces of analytic functions, like the ones in Example 1, but also to spaces of quite different kind, like the ones in Example 2. This flexibility is further underlined by the following observation.

Let us denote by $L(X_1;X_s)$ the Banach space of bounded linear operators of X_1 into X_s. The "natural" injection of X_s into $X_{s'}$, for $s' \leq s$ induces a continuous injection, also "natural," also with norm ≤ 1, of $L(X_1;X_s)$ into $L(X_1;X_{s'})$. And the composition from the left, $T \rightsquigarrow A(t)\circ T$, defines a bounded linear operator (which we denote also by $A(t)$) of $L(X_1;X_s)$ into $L(X_1;X_{s'})$ $(s' < s)$ which depends holomorphically on t, $|t| < \rho$, and whose norm is $\leq \frac{C}{s - s'}$. In other words, Properties (5) and (6) hold when $L(X_1;X_s)$ is substituted for X_s. Consequently, we may apply Th. 1 to the problem:

$$(9) \qquad (d/dt)\mathcal{R} = A(t)\mathcal{R} \ , \quad |t| < \delta \ , \quad \mathcal{R}|_{t=0} = I_1 \ ,$$

where I_1 is the identity mapping of X_1. In fact, there is advantage in translating the situation about an arbitrary point t' of the disk $|t| < \rho$, and apply Th. 1 to the problem:

$$(10) \qquad (d/dt)\mathcal{R}(t,t') = A(t)\mathcal{R}(t,t') \ , \quad |t - t'| < \rho - |t'| \ ,$$

$$(11) \qquad \mathcal{R}(t',t') = I_1 \ , \quad \text{identity of } X_1 \ .$$

The solution $\mathcal{R}(t,t')$ of (10)-(11) will be called the <u>resolvent</u> of the differential operator $L = d/dt - A(t)$. It is easily shown that it can be extended as a bounded linear operator of X_s into $X_{s'}$ for all values of s , s' such that $0 \leq s' < s \leq 1$ (and not only when $s = 1$; however the variation of t must be restricted to disks of the form $|t - t'| < \delta(s,s')$). Furthermore, $\mathcal{R}(t,t')$ is a holomorphic function of (t,t') in a set of the kind $|t - t'| < \delta(s,s')$, $|t'| < \eta < \rho$, with values in $L(X_s;X_{s'})$.

We must mention that we have the standard formulae:

$$\mathcal{R}(t_1,t_2)\mathcal{R}(t_2,t_3) = \mathcal{R}(t_1,t_3) \ ; \quad \mathcal{R}(t_1,t_2)\mathcal{R}(t_2,t_1) = \text{identity},$$

and
$$(\partial/\partial t')\mathcal{R}(t,t') = -\mathcal{R}(t,t')A(t') \ .$$

The relevance of the resolvent $\mathcal{R}(t,t')$ is clearly shown by the following:

i) The solution u of (1)-(2) can be written

(12)
$$u(t) = \mathcal{R}(t,0)u_o + \int_0^t \mathcal{R}(t,t')f(t')\,dt' \ ,$$

as immediately checked (use the uniqueness in Th. 1).

ii) Requiring now that t be real, and vary in the interval $)-\rho,\rho($, we obtain a fundamental kernel of $L = d/dt - A(t)$ by taking

(13)
$$K(t,t') = \frac{1}{2}\,\text{sgn}(t - t')\mathcal{R}(t,t') \ .$$

In fact, this is a two-sided fundamental kernel of L , for if we set

$$K\phi(t) = \int_{-\infty}^{+\infty} K(t,t')\phi(t')dt' = \frac{1}{2}\left(\int_{-\infty}^{t} - \int_{t}^{+\infty}\right)\mathcal{R}(t,t')\phi(t')\,dt' \ ,$$

for ϕ a $\mathcal{C}^\infty$ (or $\mathcal{C}^1$) function with compact support in a sufficiently small interval, valued in X_s , we have

$\mathbb{L}K\phi = K\mathbb{L}\phi = \phi$.

From the fact that $K(t,t')$ is an analytic function of (t,t') when $t \neq t'$ (in a neighborhood of $(0,0)$ in R^2), we derive at once that the differential operator $\mathbb{L}$ is hypoelliptic and analytic hypoelliptic, in the following sense:

(14) *Let $\mathcal{U}$ be an open subset of the open interval $]-\rho,\rho[$, T an arbitrary distribution in $\mathcal{U}$ valued in X_s. Suppose that $\mathbb{L}T$ is a $\mathcal{C}^\infty$ function (resp. an analytic function) in $\mathcal{U}$, valued in X_s. Then, whatever $s' < s$, T itself is a $\mathcal{C}^\infty$ (resp. analytic) function in $\mathcal{U}$, valued in $X_{s'}$.*

Another consequence of the hypoellipticity of $\mathbb{L}$ and of the preceding considerations is the existence and uniqueness of solutions in the space of distributions with support in the half-line $\{t \in R^1 ; t \geq 0\}$:

Theorem 2. Let f *be any distribution in the interval* $|t| < \rho$, *valued in* X_1, *vanishing for* $t < 0$.

There is a unique distribution u, *defined in the interval* $|t| < \delta$ *for some* $\delta < \rho$, *valued in* X_s *for some* s, $0 \leq s < 1$, *and satisfying* $Lu = f$ *for* $|t| < \delta$ *Then this is true for all* s, $0 \leq s < 1$, *provided that we take* $\delta = \delta(s) = \alpha(1 - s)$.

Further properties of the differential operator $d/dt - A(t)$, such as hypoellipticity and analytic-hypoellipticity (in an appropriate sense, allowing the "deterioration" of the values space from X_s to $X_{s'}$ with $s' < s$), are easy to establish. For the details and the proofs, see Treves [2].

3. The spaces of ultradistributions

The spaces which are going to play the role of X_s are related (via Fourier transformation) to the spaces $L^p(g)$ considered in Example 2. We shall restrict ourselves to the case $p = 2$.

Let us denote by $(\mathcal{S} \cap \mathrm{Exp})$ the space of $\mathcal{C}^\infty$ functions of the variable $x = (x_1, \cdots, x_n)$ which can be extended to C^n as entire functions of exponential type and whose derivatives of all orders decay at infinity, in the real space R^n, faster than any power of $1/|x|$. By the Paley-Wiener theorem we know that the Fourier transformation is a bijection of $(\mathcal{S} \cap \mathrm{Exp})$ onto the space $\mathcal{C}_c^\infty(R^n)$. Observing that the latter is contained in $L^2(g)$ whatever the real valued continuous function g in R^n (the fact that now g can be < 0 changes nothing to the definition of $L^2(g)$), we define on $(\mathcal{S} \cap \mathrm{Exp})$ the norm

$$\left(\int_{R_n} e^{2g(\xi)} |\hat{\phi}(\xi)|^2 d\xi \right)^{1/2},$$

where $\hat{\phi}(\xi) = \int e^{-i\langle x,\xi\rangle} \phi(x)\, dx$ is the Fourier transform of ϕ. We denote by $\mathcal{F}^{-1}L^2(g)$ the completion of $(\mathcal{S} \cap \mathrm{Exp})$ for that norm; we obtain thus a Hilbert space, and the Fourier transformation extends from $(\mathcal{S} \cap \mathrm{Exp})$ as an isometry of $\mathcal{F}^{-1}L^2(g)$ onto $L^2(g)$. All the spaces obtained in this manner are copies of L^2. Observe that

$$\int_{R^n} u(x)\overline{v(x)dx} = \int_{R^n} e^{g(\xi)}\hat{u}(\xi)\, e^{-g(\xi)}\overline{\hat{v}(\xi)}\, d\xi$$

can be extended from $(\mathcal{S} \cap \mathrm{Exp}) \times (\mathcal{S} \cap \mathrm{Exp})$ to $\mathcal{F}^{-1}L^2(g) \times \mathcal{F}^{-1}L^2(-g)$ as a continuous sesquilinear functional, and turns $\mathcal{F}^{-1}L^2(g)$ into the antidual of $\mathcal{F}^{-1}L^2(-g)$. The canonical linear isometry of $\mathcal{F}^{-1}L^2(g)$ onto its antidual identified with $\mathcal{F}^{-1}L^2(-g)$, is given by

$$u \longrightarrow \mathcal{F}^{-1}(e^{2g}\,\mathcal{F} u) \ .$$

Also observe that if g_1 is another real continuous function in R^n such that $g_1 - g \leq M$, a constant function in R^n, then $\mathcal{F}^{-1}L^2(g)$ is "contained" in $\mathcal{F}^{-1}L^2(g_1)$ and the natural injection of $\mathcal{F}^{-1}L^2(g)$ into $\mathcal{F}^{-1}L^2(g_1)$ is continuous and has norm $\leq e^M$.

The following statements are easy to check:

Proposition 1. *Suppose that* $g(\xi)/\log(1 + |\xi|)$ *tends to* $+\infty$ *as* $|\xi| \to +\infty$. *Then:*

i) $\mathcal{F}^{-1}L^2(g)$ *consists of* $\mathcal{C}^\infty$ *functions in* R^n ;

ii) *every distribution which is a finite sum of derivatives of* L^2 *functions* (in particular, every distribution with compact support) *belongs to* $\mathcal{F}^{-1}L^2(-g)$.

Proposition 2. *Suppose that* $g(\xi) \leq 0$. *Then* $\mathcal{S}$ *is contained and dense in* $\mathcal{F}^{-1}L^2(g)$.

Proposition 3. *Let* N *be a norm in* R^n, N* *the dual norm. Suppose that* $N^* \leq g +$ *a constant function in* R^n. *Then* $\mathcal{F}^{-1}L^2(g)$ *consists of functions which can be extended as holomorphic functions in the open "polycilinder" of* C^n, $\{z = x + iy \in C^n \ ; \ N(y) < 1\}$.

We shall now specify the function g. It will be the function $sh(\xi)$ where s is an arbitrary real number, and

$$h(\xi) = \sum_{j=1}^{n} (1 + \xi_j^2)^{1/2} \ . \tag{15}$$

The corresponding space $\mathcal{F}^{-1}L^2(g)$ will be denoted by K^s; K^s is a Hilbert space, a copy of L^2; $K^0 = L^2$. It follows from Prop. 3 that, when s is > 0, K^s consists of functions which can be extended as holomorphic functions in the polycilinder $\{z = x + iy \in C^n; |y_j| < s, j = 1,\cdots, n\}$. When $s < 0$, K^s contains all distributions which are finite sums of derivatives of L^2 functions; in particular, if $s < 0$, $\mathcal{E}'$, space of distributions with compact support, is contained in K^s; and also $\mathcal{S}$ is contained and dense in K^s.

The essential feature of the function $h(\xi)$, (15), is that towards infinity, it behaves like $|\xi_1| + \cdots + |\xi_n|$. Because of the separation of variables in $H(\xi)$, the proofs of the statement below can be reduced to the case of a single variable, i.e., the case $n = 1$. Let us set, for $j = 1, \cdots, n$ (R is any number > 0),

$$X_j = \tan^{-1}(x_j/R), \quad Y_j = x_j/(x_j^2 + R^2), \quad Z_j = (x_j^2 + R^2)^{-1} .$$

Proposition 4. *For all real* $s < R$, *the functions* $Y_1\cdots Y_n$ *and* $Z_1\cdots Z_n$ *belong to* K^s.

Proposition 5. *There is a continuous function* $C(R,s)$ *of* (R,s), $R > 0$, s *real*, $|s| < R$, *such that, if* p, q, r *are three arbitrary n-tuples, multiplication by*

$$X_1^{p_1} \cdots X_n^{p_n} Y_1^{q_1} \cdots Y_n^{q_n} Z_1^{r_1} \cdots Z_n^{r_n}$$

defines a bounded linear operator of K^s *into itself with norm* $C(R,s)^{|p+q+r|}$.

As we have said, it is enough to consider the case $n = 1$ (then we denote the variable by x and ξ). Observe that the Fourier transform of the function $(x + iR)^{-1}$ is

$$- 2i\pi Y(\xi)\, e^{-R|\xi|} ,$$

where $Y(\xi)$ is the Heaviside function, $Y(\xi) = 1$ for $\xi > 0$, $Y(\xi) = 0$ for $\xi < 0$. It follows at once (in the case $n = 1$) that $(x + iR)^{-1}$ and $(x - iR)^{-1}$ belong to K^s if $s < R$, and define multipliers on K^s if $|s| < R$. As a matter of fact, the norm of those multipliers is $\leq (R - s)^{-1}$. But we have:

$$Y = (x^2 + R^2)^{-1}x = (1/2)\,\{(x + iR)^{-1} + (x - iR)^{-1}\}\ ,$$

$$Z = (x^2 + R^2)^{-1} = (1/2R)\,\{(x - iR)^{-1} - (x + iR)^{-1}\}\ ,$$

whence Prop. 4 and the portion of Prop. 5 which does not involve the X_j's. When these intervene, we observe that $\mathfrak{F}(\tan^{-1}(x/R)) = i\pi e^{-R|\xi|}\mathrm{pv}(1/\xi)$ is a convolver on $L^2(s\sqrt{1 + \xi^2})$ for $|s| < R$.

These results will be useful in the next section. Let us say right away that, in applying Th. 1, we shall take, as spaces X_s, tensor products $K^s \otimes E$ with E a finite dimensional Hilbert space. The fact that s varies over the whole real line, instead of the unit interval, is of no importance: we shall always restrict the variation of s to a compact interval, not reduced to a point, and we can transform such an interval into the unit one by an increasing diffeomorphism. This would not affect the basic properties (5) and (6). That (6) holds in the situation which we shall study, will follow from the next statement:

<u>Proposition 6</u>. <u>Whatever the real numbers</u> s, s', $s < s'$, <u>the partial differentiations</u> $\partial/\partial x_j (1 \leq j \leq n)$ <u>define bounded linear operators</u> $K^s \to K^{s'}$ <u>with norm</u> $\leq e^{-1}(s - s')^{-1}$.

<u>Proof</u>: $\mathfrak{F}((\partial/\partial x_j)u) = i\,\xi_j \mathfrak{F}u$ and $|\xi_j|\ \exp(-(s-s')h(\xi)) \leq e^{-1}(s - s')^{-1}$ where h is given by (15).

4. The change of variables

We now go back to the differential operator P and the piece of analytic hypersurface S. By shrinking, if necessary, the open set $\mathcal{O}$ where P and S are defined, we may find an analytic change of variables which transforms S into a piece of hyperplane. Let $(y_1, \cdots, y_n, y_{n+1})$ by the new coordinates (we shall systematically write $y = (y_1, \cdots, y_n)$) and suppose that $y_{n+1} = 0$ is the equation of S in $\mathcal{O}$. The transformation we are now about to perform, and its subsequent exploitation apply to arbitrary, but <u>determined</u>, systems of linear partial differential operators with analytic coefficients. However, for the sake of simplicity, we shall restrict ourselves to the case of a system of first order differential operators. For the higher order case, either one reduces it to first order by adjoining additional unknowns (this is always feasible) or else modifies slightly the technique (see Treves [1], [2]). After perhaps some further shrinking of $\mathcal{O}$, using the fact that S is nowhere characteristic with respect to P, we may write

$$(16) \qquad P = D_{y_{n+1}} + \sum_{j=1}^{n} \alpha_j(y,y_{n+1})\, D_{y_j} + \alpha_0(y,y_{n+1}),$$

where $D_{y_j} = \partial/\partial y_j$ $(1 \leq j \leq n+1)$ and where the α_j $(0 \leq j \leq n)$ are $N \times N$ matrices with complex valued analytic functions as entries. Next we select δ, $\varepsilon > 0$ sufficiently small so that the closure of the set Ω now defined:

$$|y_j| < \delta\pi/2, \ \ j = 1, \cdots, n; \ |y_{n+1}| < \varepsilon\Big(\prod_{j=1}^{n} \cos^2(y_j/\delta)\Big) ,$$

be contained in $\mathcal{O}$. For (y,y_{n+1}) in Ω we set

$$(17) \qquad y_j = \delta \tan^{-1}(x_j/R),\ j = 1, \cdots, n;\ y_{n+1} = \varepsilon t \prod_{j=1}^{n} (x_j^2 + R^2)^{-1}.$$

With the notation used in Prop. 4 and Prop. 5, this can be rewritten:

$$y_j = \delta X_j, \quad j=1,\cdots,n; \quad y_n = \varepsilon t Z_1\cdots Z_n.$$

The motivation for such change of variables will soon become apparent. At any rate it defines a proper analytic diffeomorphism between Ω and the unit strip $\Sigma = \{(x,t) \in R^{n+1}; \ |t| < 1\}$. We have

$$\delta D_{y_j} = Z_j^{-1}(R\, D_{x_j} + 2tY_j\, D_t)\ (1 \leq j \leq n), \quad D_{y_{n+1}} = (\varepsilon Z_1\cdots Z_n)^{-1}D_t$$

Let us call $Q = Q(x,t,D_x,D_t)$ the transform of P (given by (16)) under the diffeomorphism (17):

$$Q = \beta_{n+1}(x,t)D_t + \sum_{j=1}^{n} \beta_j(x,t)\, D_{x_j} + \beta_0(x,t) \ .$$

We have $\beta_0(x,t) = \alpha_0(\delta X,\varepsilon t Z_1\cdots Z_n)$ and

$$\beta_j(x,t) = R(\delta Z_j)^{-1}\, \alpha_j(\delta X,\varepsilon t Z_1\cdots Z_n) \ , \qquad j=1,\cdots,n;$$

$$\beta_{n+1}(x,t) = (\varepsilon Z_1\cdots Z_n)^{-1}\{1 + 2\,\frac{\varepsilon}{\delta} \sum_{j=1}^{n} (\prod_{\substack{k=1\\k\neq j}}^{n} Z_k)\, Y_j\, \alpha_j(\delta X,\varepsilon t Z_1\cdots Z_n)\} \ ,$$

where X stands for the vector $(X_1,\cdots,X_n)$. Now we use the analyticity of the coefficients $\alpha_j(y,y_{n+1})$. We assume that whatever $j=0,\cdots,n$, the radiuses of convergence (with respect to $y_1,\cdots,y_{n+1}$) of the Taylor expansion about the origin of α_j are $>$ a given number $\rho > 0$. Let now σ be a number such that $0 < \sigma < R$. We apply Prop. 5 with $|s| \leq \sigma$. We see at once that if δ, ε and ε/δ are sufficiently small, multiplication by $\beta_j(x,t)$ $(0 \leq j \leq n)$ and by $\beta_{n+1}(x,t)^{-1}$ defines bounded linear operators of $K^s \otimes C^N$ into itself. We set

$$L = D_t + \sum_{j=1}^{n} \gamma_j(x,t)D_{x_j} + \gamma_0(x,t) ,$$

where $\gamma_j(x,t) = \beta_{n+1}(x,t)^{-1}\, \beta_j(x,t)$.

Suppose that we wanted to solve the Cauchy problem:

(18) $$Qu = f \quad \text{for} \quad |t| < \eta, \quad u\big|_{t=0} = u_o ,$$

with data f, u_o valued in some space $K^{s_o} \otimes C^N$ ($|s_o| \leqslant \sigma$). It will be equivalent to solve the following Cauchy problem:

(19) $$Lu = \beta_{n+1}^{-1} f, \quad |t| < \eta, \quad u\big|_{t=0} = u_o .$$

For observe that $\beta_{n+1}^{-1} f$ is also a function of t valued in $K^{s_o} \otimes C^N$, holomorphic (resp. real analytic, resp. $\mathcal{C}^k$) if this is true of f.

5. The main result

We consider the Cauchy problem (19). Then, if we keep in mind the fact that the coefficients of L, the $\gamma_j(x,t)$ $(0 \leqslant j \leqslant n)$, are multipliers on $K^s \otimes C^N$ for $|s| \leqslant \sigma$, depending "holomorphically" on t, and if we take advantage of Prop. 6, we see at once that

(20) $$A(t) = - \sum_{j=1}^{n} \gamma_j(x,t)D_{x_j} - \gamma_0(x,t)$$

fulfills Condition (6), and that we may therefore apply Th. 1 with $X_s = K^s \otimes C^N$ and $-\sigma \leqslant s \leqslant s_o$. We shall state both the "real" and the "complex" version of this particular case of Th. 1:

<u>Theorem 3</u>. <u>Let</u> s_o <u>be a real number</u>, $|s_o| < \sigma$. <u>Let</u> $f(t)$ <u>be a continuous</u> (<u>resp</u>. <u>holomorphic</u>) <u>function of the real</u> (<u>resp</u>. <u>complex</u>) <u>variable</u>

t, $|t| < \rho$ valued in $K^{s_o} \otimes C^N$, and let u_o be an arbitrary member of $K^{s_o} \otimes C^N$.

There is a unique solution $u(t)$ of (18) which for some $\eta > 0$ and some s, $-\sigma \leqslant s < s_o$, is a continuously differentialbe function of t, $|t| < \rho$, valued in $K^s \otimes C^N$. Then, for some constant $\alpha > 0$ and all s, $-\sigma \leqslant s < s_o$, $u(t)$ is a $\mathcal{C}^1$ function of t, $|t| < \alpha(1 - s)$, valued in $K^s \otimes C^N$.

In the complex case, "continuously differentiable" and $\mathcal{C}^1$ mean "holomorphic."

When the variable t is real, we have also the analog of Th. 2, that is, the distributions version of Th. 3. Its statement is obvious. Let us indicate two simple applications of these results:

1) The diffeomorphism (17) transforms the holomorphic functions of complex t, $|t| < 1$, valued in K^s, $s > 0$, into analytic functions in the set Ω. Conversely, every function which can be extended as a holomorphic function in an open neighborhood of $\bar{\Omega}$ in C^{n+1} corresponds, via (17), to a holomorphic function of t, $|t| < \rho$ for some $\rho > 0$, valued in K^s for some $s > 0$. In view of this, by choosing $s_o > 0$ in Th. 3, we obtain the Cauchy-Kovalevska theorem for systems of first order linear partial differential operators with analytic coefficients (it is well known that the general linear Cauchy Kovalevska theorem can be reduced to this case by adjoining equations and unknown to the system under study).

2) Let us go back to the original situation $\mathcal{O}$, P, S. Let u be a (vector valued) distribution in $\mathcal{O}$ satisfying $Pu = 0$ and whose support lies on one side of S, say in the region $y_{n+1} \leqslant 0$. Suppose that for

$d > 0$ and tending to 0, the diameter of the set (supp u) $\cap$ $\{(y, y_{n+1});$ $-d \leq y_{n+1} \leq 0\}$ tends to zero. Suppose for a moment that the origin belongs to supp u. Let us transfer the restriction of u to Ω, to the unit strip $\{(x,t) \in R^{n+1};\ |t| < 1\}$, by means of (17). The transform v of u satisfies $Qv = 0$ in this strip and vanishes for $t > 0$. Furthermore, v is a distribution with respect to t with values in $\mathcal{D}'_x \otimes C^N$; but for $|t|$ small enough, these values are distributions in the space variables x having a compact support. Consequently, for $|t| < \eta$ small enough, $v(t)$ is a distribution of t with values in $K^s \otimes C^N$ whatever $s > 0$. By the uniqueness in the distribution version of Th. 3 (cf. Th. 2), v must vanish for $|t| < \eta$. Reverting to u, we see the origin cannot belong to supp u. Repeating this reasoning for every point of S, we see that u must vanish in a neighborhood of S. This is the generalized Holmgren's theorem (see Hörmander [1], Th. 5.3.1).

3) Let now f be a vector valued continuous function with compact support in Ω, u_o a vector-valued continuous function of $y = (y_1, \cdots, y_n)$, having a compact support contained in the open set $\{y \in R^n;\ (y,0) \in \Omega\}$. Consider the problem $Pu = f$, $u|_{y_{n+1}=0} = u_o$ (P given by (16)). We may transform it into the problem (18) and apply Th. 3. We reach thus the conclusion that it has a unique solution. But the nature of this solution raises some questions. In general, it is not going to be a distribution in Ω. The diffeomorphism (17) transforms the space of K^s-valued $(s \geq 0)$ $\mathcal{C}^\infty$ functions of t, $|t| < 1$, onto a space of functions in Ω. There is a transpose to this transformation, associated with (17), which induces an isomorphism between the dual of this space of functions in Ω and the space of distributions of t, $|t| < 1$, valued in K^{-s}. The

solution u of our Cauchy problem belongs to this dual. If we need more information about it, one way of getting some is to look at its transform in the unit strip $|t| < 1$. This is a $\mathcal{C}^1$ function of t, and its Fourier transform with respect to x is a <u>function</u> of ξ. Another more direct, although probably less explicit, approach is through the theory of analytic functionals. As a matter of fact, a direct (that is, avoiding changes of variables of the type of (17)) proof of the existence and uniqueness of the solution u is possible if we combine the theory of analytic functionals with Th. 1. For the details, see Treves [2], Sections 6 and 11. But let us emphasize that the solution u to the Cauchy problem for P is a distribution if and only if this is true about the solution to the transformed Cauchy problem, i.e., the Cauchy problem relative to Q. For the diffeomorphism (17) transforms distributions into distributions, and $\mathcal{C}^k$ (resp. analytic, etc.) functions into $\mathcal{C}^k$ (resp. analytic, etc.) functions. If therefore we wish to study the question as to whether the solution to the Cauchy problem relative to P, with data f, u_0 in spaces of functions or distributions with compact support, is itself a distribution (or a function with a certain degree of regularity), we might as well study the same question for the operator Q or, even better, for the operator L. We may then try to take full advantage of the Fourier transformation with respect to the space variables x, since now the problem has been "globalized" with respect to these variables. The next and last section deals briefly with this important question.

6. <u>The symbol of the resolvent</u>

We shall only be concerned here with the differential operator $d/dt - A(t)$, where $A(t)$ is given by (20), and with its resolvent $\mathcal{R}(t,t')$

(see Eqq. (10) - (11)). Let ξ be an arbitrary point of R^n and consider the following Cauchy problem:

(21) $$(\partial/\partial t)u = A(t)u, \qquad |t - t'| < \eta,$$

(22) $$u\big|_{t=t'} = Ie^{i<x,\xi>},$$

where I stands for the $N \times N$ identity matrix. Here we look for a solution $u(x,t)$ which is a function of x in R^n, of t in some neighborhood of t', with values in the space $M_N(C)$ of $N \times N$ matrices with complex entries. This solution u will depend on t' and on ξ, so that it is appropriate to set $u = u(x, \xi; t, t')$. If we wish to prove the existence and uniqueness of u, also that it depends continuously on t' and on ξ (which is important for what we want to do with u), we cannot follow blindly the method used in the earlier sections and content ourselves with substituting everywhere $M_N(C)$ for C^N. The reason we cannot do so is that the function of x, $\exp(i<x,\xi>)$, never belongs to K^s (whatever the real number s). A slight modification of our choice of spaces X_s will, however, do the trick. We observe that, given any $R > 0$,

$$Z_1 \cdots Z_n\, e^{i<x,\xi>} \qquad \text{(see notation in p. 14)}$$

belongs to K^s for all $s < R$ (if we take Prop. 4 into account, this is simply saying that $\exp(i<x,\xi>)$ is a multiplier on K^s for all s, which is evident). Whence, quite naturally, the idea of introducing the space, which will be denoted by ${}^1K^s$, of ultradistributions u such that $Z_1 \cdots Z_n u$ belongs to K^s. This can be defined rigourously; for the details, see Treves [2], Ch. 3. Now we shall apply Th. 1 taking, as spaces X_s

the tensor products ${}^{1}K^{s} \otimes M_n(C)$. Our results about the operator $A(t)$ and the spaces K^s yields at once that Conditions (5) and (6) are, now also, satisfied. The existence and uniqueness of the solution u to (21)-(22) follows at once. As a matter of fact we have

$$u = u(x,\xi;t,t') = \mathcal{R}(t,t')(Ie^{i<x,\xi>}) .$$

One derives easily from this that u is holomorphic with respect to t, t' in a suitable neighborhood of the origin in C^2, is analytic in x and can be extended as a holomorphic function in the polycilinder $|\mathrm{Im}\, x_j| < R$, $j=1,\cdots,n$, and is continuous with respect to ξ in the whole of R^n. Furthermore, we have, for $0 < s < R$, (t,t') in a suitable neighborhood of 0 in C^2 (depending on s, more specifically on s^{-1} and on $(R - s)^{-1}$), all x and ξ in R^n,

$$\|u(x,\xi;t,t')\| \leq \text{const.} \prod_{j=1}^{n} (R^2 + x_j^2)\, s^{-1}\, e^{sh(\xi)} ,$$

where $\|\ \|$ is a norm on $M_n(C)$. Let us set

$$\Upsilon(x,\xi;t,t') = e^{-i<x,\xi>}\, u(x,\xi;t,t') .$$

We may say that $\Upsilon(x,\xi;t,t')$ is the <u>symbol</u> of the resolvent $\mathcal{R}(t,t')$. For let ϕ be a vector-valued function in R^n whose Fourier transform $\hat{\phi}$ is $\mathcal{C}^\infty$ and has compact support. We have

(23) $$\mathcal{R}(t,t')\phi = (2\pi)^{-n}\int_{R_n} e^{i<x,\xi>}\, \Upsilon(x,\xi;t,t')\hat{\phi}(\xi)\, d\xi .$$

Indeed, observe that the right hand side satisfies the homogeneous equation $\{(\partial/\partial t) - A(t)\}. = 0$, and that for $t = t'$, it is equal to $\phi(x)$. Therefore, in view of (12), we must have (23).

I wish to indicate briefly what use can be made of an expression such as (23). For simplicity, I shall restrict myself to the homogeneous Cauchy problem: this is the case where the right hand side f in Eq. (1) is identically equal to zero. The inhomogeneous Cauchy problem raises some questions which would carry us too far. We shall only deal with differential operators P of the form (16), as we have described here the transformation only for such a form; it is not difficult, however, to perform it on general determined systems as we have already said. Let us go back to the open set $\mathcal{O}$, which we assume to be given by inequalities $|y_j| < r_j$ $(1 \leq j \leq n+1)$, and to the piece of hypersurface S in $\mathcal{O}$, given by $y_{n+1} = 0$. We set

$$\mathcal{O}' = \{y \in R^n; \ |y_j| < r_j, \ j = 1, \cdots, n\} .$$

We shall say that the homogeneous local Cauchy problem, relative to P and to S at the origin, is well-posed if we can choose the numbers $r_j > 0$ $(1 \leq j \leq n+1)$ in such a way that the following holds:

(24) For every vector-valued $\mathcal{C}^\infty$ function $u_o(y)$ with compact support in $\mathcal{O}'$ there is a unique solution $u(y,y_{n+1})$ of

$$Pu = 0 \ \text{in } \mathcal{O}, \quad u(y,0) = u_o(y) \ \text{in } \mathcal{O}' ,$$

which is a continuous function of y_{n+1}, $|y_{n+1}| < r_{n+1}$, valued in the space of (vector-valued) distributions in $y \in \mathcal{O}'$.

Consider then $v_o(x)$ an arbitrary element of $\mathcal{S} \otimes C^N$, i.e., a $\mathcal{C}^\infty$ function of x which decays at infinity, together with all its derivatives, faster than any power of $1/|x|$. We set

$$u_o(y) = v_o(R\tan(y/\delta)) \qquad (\text{cf. } (17), \quad \tan\frac{y}{\delta} = (\tan\frac{y_1}{\delta}, \cdots, \tan\frac{y_n}{\delta}))$$

It is easy to see that $u_o(y)$ is a $\mathcal{C}^\infty$ function in $\Omega' = \{y \in R^n;\ |y_j| < \delta/2$ $j = 1, \cdots, n\}$, which can be extended to R^n as a $\mathcal{C}^\infty$ function with support in $\overline{\Omega'}$. As before, we choose $\delta > 0$ so small that $\overline{\Omega'}$ is a compact subset of $\mathcal{O}'$. Let $u(y,y_{n+1})$ be the distribution whose existence (and uniqueness) is stated in (24). By using the diffeomorphism (17), we can pull back into R^n its restriction to Ω: we obtain thus a distribution $v(x,t)$ which, as one can easily see, is a continuous function of t, $|t| < 1$, with values in space $\mathcal{S}'_x \times C^N$ of tempered distributions in the space variables x. We must have

$$v(x,t) = \mathcal{K}(t,0)v_o = (2\pi)^{-n}\int_{R^n} e^{i<x,\xi>}\Gamma(x,\xi;t,0)\ \hat{v}_o(\xi)\ d\xi\ .$$

Recalling that v_o, and therefore $\hat{v}_o$, is arbitrary in $\mathcal{S} \otimes C^N$, we apply The Schwartz kernel theorem and conclude easily that

(25) $\Gamma(x,\xi;t,0)$ <u>is a continuous function of</u> t, $|t| < 1$, <u>valued in the space</u> $\mathcal{S}'_{x,\xi} \times M_N(C)$ <u>of tempered distribution in</u> $(x,\xi) \in R^n \times R^n$ <u>with values in the ring of</u> $N \times N$ <u>complex matrices</u> (see Schwartz [1]).

Conversely, by reverting the previous argument, it is not difficult to check that one can deduce (24) from (25), provided that one takes now the numbers r_j $(1 \leq j \leq n+1)$ sufficiently small (the set $\mathcal{O}$ to which we come back in this manner will be contained in the set Ω, whereas before the opposite was true).

The equivalence of (24) and (25), in the sense outlined above, emphasizes the importance of finding out when does (25) hold. Sufficient conditions are known, although they are certainly far from being necessary. When the coefficients of the differential operator $A(t)$, given by (20), that is, the matrices $\gamma_j (0 \leq j \leq n)$ are constant, a complete answer, that is conditions which are both necessary and sufficient, are provided by Gårding's theorem ([1] see also Friedman [1], Ch. 7). Consider the matrix

$$\gamma(\xi) = - \sum_{j=1}^{n} \gamma_j \xi_j + i\gamma_o .$$

In this case, the symbol of the resolvent is an exponential:

$$\Gamma(\xi;t,t') = e^{i(t-t')\gamma(\xi)} .$$

It is not difficult see (cf. Friedman [1], Ch. 7, Th. 3) that

(26) $$\| \Gamma(\xi;t,0) \| \leq \text{const.}\ (1 + |t\xi|)^{n-1} e^{|t|\Lambda(\xi)} ,$$

where $\Lambda(\xi)$ is the supremum of the quantities $\mathrm{Im}\lambda_j(\xi)$, where $\lambda_j(\xi)$ ranges over the set of eigenvalues of the matrix $\gamma(\xi)$. It is evident in (26) that $\Gamma(\xi;t,0)$ will be of slow growth at infinity, with respect to ξ, i.e., will belong to $(\mathcal{O}_M)_\xi \otimes M_N(C)$, if

(27) $$\sup_{\xi \in R^n} \Lambda(\xi) < +\infty.$$

By applying the Seidenberg-Tarski theorem, one can show in fact that (27) is necessary in order that $\Gamma(\xi;t,0)$ have a slow growth at infinity in ξ (observe that

$$e^{t\Lambda(\xi)} \leq \| \Gamma(\xi;t,0) \|) .$$

If (27) holds, one says that the differential operator $\frac{\partial}{\partial t} + \sum_{j=1}^{n} \gamma_j \frac{\partial}{\partial x_j} + \gamma_o$ is <u>hyperbolic</u> in the sense of Gårding. We should say that, in this case, not only is the local homogeneous Cauchy problem well-posed, but also the inhomogeneous one: one can see this at once by applying Formula (12) and using the fact that $\Gamma(\xi;t,t') = \Gamma(\xi;t-t',0)$.

When the coefficients are variable, that is, $\gamma_j = \gamma_j(x,t)$, there is an important case where the local Cauchy problem, homogeneous as well as inhomogeneous, is well posed. This is when the eigenvalues of the matrix

$$-\sum_{j=1}^{n} \gamma_j(x,t)\xi_j$$

are <u>real</u> and <u>distinct</u> whatever $\xi \neq 0$ in R^n. In this case, one says that the system of differential operators is <u>strictly hyperbolic</u>. Of course, when the coefficients are constant, strictly hyperbolic implies hyperbolic in the sense of Gårding. One can easily check this directly. The converse is not true. Outside of the strictly hyperbolic case, nothing general is known in the variable case, and important problems remain to be solved in this connection.

References

A. Friedman [1]. *Generalized functions and partial differential equations*, Prentice-Hall Englewood Cliffs N.J. 1963.

L. Garding [1]. *Linear hyperbolic partial differential equations with constant coefficients*. Acta Math. 85, 1-62 (1950).

L. Hörmander [1]. *Linear partial differential operators*, Springer Berlin 1963.

H. Lewy [1]. *An example of a smooth linear partial differential equation without solution*. Ann. Math. (2), 66, 155-158 (1957).

L. Nirenberg & F. Treves [1]. *Solvability of a first order partial differential equation*, Comm. Pure Appl. Math. 16, 331-351 (1963).

L.V. Ovciannikov [1]. Singular operators in Banach spaces scales (Russian) Doklady Akad. Nauk. 163, n°4, 819-822 (1965.

P. Schapira [1]. *Une equation aux dérivées partielles sans solutions dans l'espace des hyperfunctions*. C. R. Acad. Sc. Paris, 265, 665-667 (1967).

L. Schwartz [1]. *Les equations d'évolution liées au produit de convolution*, Ann. Inst. Fourier 2, 19-49 (1950-51).

F. Treves [1]. *On the theory of linear partial differential operators with analytic coefficients*. Trans. Amer. Math. Soc. (to appear).

F. Treves [2]. *Ovciannikov theorem and applications*, to appear.

<u>Footnote</u> p. 75

(*) After this lecture was delivered, I have learned that the main theorem of this section, Th. 1, had already been proved by L. V. Ovciannikov in 1965 (see Ovciannikov [1]).

<u>Footnote</u> p. 78

(*) See Ovciannikov [1].

Algebraic Topology and Operators in Hilbert Space

by M. F. Atiyah*

Introduction.

In recent years considerable progress has been made in the global theory of elliptic equations. This has been essentially of a topological character and it has brought to light some very interesting connections between the topology and the analysis. I want to talk here about a simple but significant aspect of this connection, namely the relation between the index of Fredholm operators and the theory of vector bundles.

§1. Fredholm operators.

Let H be a separable complex Hilbert space. A bounded linear operator

$$F : H \longrightarrow H$$

is called a *Fredholm operator* if its kernel and cokernel are both finite-dimensional. In other words the homogeneous equation $Fu = 0$ has a finite number of linearly independent solutions, and the inhomogeneous equation $Fu = v$ can be solved provided v satisfies a finite number of linear conditions. Such operators occur frequently in various branches of analysis particularly in connection with elliptic problems. A useful criterion characterizing Fredholm operators is the following

PROPOSITION 1.1. *F is a Fredholm operator if and only if it is invertible modulo compact operators, i.e. $\exists G$ with $FG-1$ and $GF-1$ both compact.*

This proposition is essentially a reformulation of the classical result of Fredholm that, if K is compact, $1+K$ is a Fredholm operator. In the theory of elliptic differential equations the approximate inverse G

*Supported in part by Air Force Office of Scientific Research grant AF-AFOSR-359-66.

is called a "parametrix."

To reformulate (1.1) in geometrical terms let $\mathcal{A}$ denote the Banach algebra of all bounded operators on H. The compact operators form a closed 2-sided ideal $\mathcal{K}$ of $\mathcal{A}$ so that $\mathcal{B} = \mathcal{A}/\mathcal{K}$ is again a Banach algebra. Let $\mathcal{B}^*$ denote the group of invertible elements in $\mathcal{B}$. Then (1.1) asserts that the space $\mathcal{F}$ of Fredholm operators is just the inverse image $\pi^{-1}(\mathcal{B}^*)$ under the natural map $\pi : \mathcal{A} \longrightarrow \mathcal{B}$. In particular this shows that $\mathcal{F}$ is an open set in $\mathcal{A}$, closed under composition, taking adjoints and adding compact operators.

The <u>index</u> of a Fredholm operator F is defined by

$$\text{index } F = \dim \operatorname{Ker} F - \dim \operatorname{Coker} F \ .$$

<u>Remark</u>. A simple application of the closed-graph theorem (see Lemma (2.1)) shows that a Fredholm operator has a closed range. Thus

$$\operatorname{Coker} F = H/F(H) \cong \operatorname{Ker} F^*$$

where F^* is the adjoint of F. Hence

$$\text{index } F = \dim \operatorname{Ker} F - \dim \operatorname{Ker} F^* \ .$$

As we have observed the composition $F' \circ F$ of two Fredholm operators is again Fredholm. A simple algebraic argument shows at once that

$$\text{index } F' \circ F = \text{index } F' + \text{index } F \ .$$

Suppose now that $\{e_0, e_1, \ldots\}$ is an orthonormal base for H and let H_n denote the closed subspace spanned by the e_i with $i \geq n$. The orthogonal projection P_n onto H_n is then Fredholm and, since it is self-adjoint, it has index zero. Thus $F_n = P_n \circ F$ is Fredholm and has the same index as F. Now choose n so that $e_0, e_1, \ldots, e_{n-1}$ and F(H) span H -- which is possible because dim $H/F(H)$ is finite. Then $F_n(H) = H_n$ and so $\operatorname{Ker} F_n^* = H_n^{\perp}$. Thus

(1.2) $$\text{index } F = \text{index } F_n = \dim \operatorname{Ker} F_n - n .$$

This is a convenient way of calculating the index because we have fixed the dimension of the cokernel to be n, and so we have only one unknown dimension to compute, namely $\dim \operatorname{Ker} F_n$.

§2. Fredholm families.

We want now to investigate families of Fredholm operators depending continuously on some parameter. Formally this means that we have a topological space X (the space of parameters) and a continuous map

$$F : X \longrightarrow \mathcal{F}$$

where $\mathcal{F}$ is topologized as a subspace of the bounded operators $\mathcal{Q}$ (with the norm topology). Thus for each $x \in X$ we have a Fredholm operator $F(x)$ and

$$\| F(x) - F(x_0) \| < \varepsilon$$

for x sufficiently close to x_0.

We now ask how the vector spaces $\operatorname{Ker} F(x)$ and $\operatorname{Ker} F(x)^*$ vary with x. It is easy to see that their dimensions are not continuous (i.e. locally constant) functions of x: they are only semi-continuous, that is

$$\dim \operatorname{Ker} F(x_0) \geq \dim \operatorname{Ker} F(x)$$

for all x sufficiently close to x_0. On the other hand, as we shall see below in (2.1), index $F(x)$ is locally constant. Thus, although $\dim \operatorname{Ker} F(x)$ and $\dim \operatorname{Ker} F(x)^*$ can jump as $x \longrightarrow x_0$, they always jump by the same amount so that their difference remains constant. This invariance of the index under perturbation is its most significant property: it brings it into the realm of algebraic topology.

To go beyond questions of dimension it will be convenient to introduce the operators

$$F_n(x) = P_n \circ F(x)$$

defined as in §1. For large n this has the effect of fixing the dimension of the spaces $\operatorname{Ker} F_n(x)$. This is a necessary preliminary if we want these spaces to vary reasonably with x. More precisely we have the following result on the local continuity properties of these kernels:

LEMMA (2.1). _Let_ $F : X \longrightarrow \mathcal{F}$ _be a continuous family of Fredholm operators and let_ $x_0 \in X$. _Choose_ n _so that_ $F_n(x_0)(H) = H_n$. _Then there exists a neighborhood_ U _of_ x_0 _so that, for all_ $x \in U$,

$$F_n(x)(H) = H_n \quad .$$

Moreover $\dim \operatorname{Ker} F_n(x)$ _is then a constant_ d _(for_ $x \in U$) _and we can find_ d _continuous functions_

$$s_i : X \longrightarrow H$$

such that, for $x \in U$, $s_1(x), \ldots, s_d(x)$ _is a basis of_ $\operatorname{Ker} F_n(x)$.

The proof of the lemma is shorter than its enunciation. Let $S = \operatorname{Ker} F_n(x_0)$ and define an operator

$$G(x) : H \longrightarrow H_n \oplus S$$

by $G(x)u = (F_n(x)u, P_S u)$ where P_S is the projection on S. Clearly $G(x_0)$ is an isomorphism.* Since $G(x)$ is continuous in x it follows that† $G(x)$ is an isomorphism for all x in some neighborhood U of x_0. This proves the lemma. In fact we clearly have $F_n(x)(H) = H_n$ for $x \in U$ and, if $e_1, \ldots, e_d$ is a basis of S,

---

*Algebraically and hence, by the closed-graph theorem, topologically.

†Modulo an identification of $H_n \oplus S$ with H this is just the assertion that the invertible elements in the Banach algebra $\mathcal{A}$ form an open set.

$$s_i(x) = G(x)^{-1}e_i \qquad i \leq i \leq d$$

give a basis for Ker $F_n(x)$.

In view of (1.2) Lemma (2.1) shows at once that index $F(x) = d-n$ is a locally constant function of x. However it gives more information which we proceed to exploit. Suppose now that our parameter space X is <u>compact</u>. The first part of (2.1), combined with the compactness of X, implies that we can find an integer n so that

$$F_n(x)H = H_n \quad \text{for all} \quad x \in X \ .$$

Choose such an integer and consider the family of vector spaces $S(x) = \text{Ker } F_n(x)$ for $x \in X$. If we topologize $S = \bigcup_{x \in X} S(x)$ in the natural way as a subspace of $X \times H$ the last part of (2.1) implies that S is a <u>locally trivial</u> family of d-dimensional vector spaces, i.e. in a neighborhood of any point x_0, we can find d continuous maps $s_i : X \longrightarrow S$ such that $s_1(x), \ldots, s_d(x)$ lie in $S(x)$ and form a basis.

A locally trivial family of vector spaces parametrized by X is called a <u>vector bundle</u> over X. Thus a vector bundle consists of a topological space S mapped continuously by a map π onto X so that each 'fibre' $S(x) = \pi^{-1}(x)$ has a vector space structure and locally we can find continuous bases $s_1(x), \ldots, s_d(x)$ as above.

There is† a natural notion of isomorphism for vector bundles over X. Let us denote by Vect(X) the set of isomorphism classes of all vector bundles over X. If X is a point a vector bundle over X is just a single vector space and Vect(X) is the set of non-negative integers $\mathbb{Z}^+$. For general spaces X however one can give simple examples of non-trivial vector bundles so that the isomorphism class of a vector bundle is not determined by its dimension.

Given two vector bundles S, T over the same space X one can form their direct sum $S \oplus T$. This is again a vector bundle over X and

---

† For definitions and elementary properties of vector bundles see [1, Chapter I].

the fibre of $S \oplus T$ at x is just $S(x) \oplus T(x)$. This induces an abelian semi-group structure on Vect(X), generalizing the semi-group structure on $\mathbb{Z}^+$.

Let us return now to the consequences of (2.1). We have seen that, given the Fredholm family

$$F : X \longrightarrow \mathcal{F}$$

(with X compact), we can choose an integer n so that $\bigcup_{x \in X} \operatorname{Ker} F_n(x)$ is a vector bundle over X. We denote this vector bundle by $\operatorname{Ker} F_n$. Moreover $\operatorname{Ker} F_n^*(x) = H_n^{\perp}$ for all x so that $\operatorname{Ker} F_n^*$ is the trivial bundle $X \times H_n^{\perp}$. In view of formula (1.2) for the index of a single Fredholm operator it is now rather natural to try to define a more general notion of index for a Fredholm family F by putting

$$\begin{aligned} \text{index } F &= [\operatorname{Ker} F_n] - [\operatorname{Ker} F_n^*] \\ &= [\operatorname{Ker} F_n] - [X \times H_n^{\perp}] \end{aligned} \tag{2.2}$$

where [ ] denotes the isomorphism class in Vect(X). Unfortunately this does not quite make sense because Vect(X), like $\mathbb{Z}^+$, is only a semi-group and not a group so that subtraction is not admissible. However the way out is fairly clear: we must generalize the construction of passing from the semi-group $\mathbb{Z}^+$ to the group $\mathbb{Z}$ and associate to the semi-group Vect(X) an abelian group K(X). There is in fact a routine construction which starts from an abelian semi-group A and produces an abelian group B. One way is to define B to consist of all pairs (a_1, a_2) with $a_i \in A$ modulo the equivalence relation generated by

$$(a_1, a_2) \sim (a_1 + a, a_2 + a) \qquad a \in A$$

(we think of (a_1, a_2) as the difference $a_1 - a_2$). The only point to watch is that the natural map $A \longrightarrow B$ need not be injective.

Having introduced our group K(X) formula (2.2) now makes sense and defines the index of a Fredholm family as an element of K(X). In this definition we had to choose a sufficiently large integer n. However if we replace n by n+1 it is easy to see that

$$\operatorname{Ker} F_{n+1} \cong \operatorname{Ker} F_n \oplus E_n$$
$$\operatorname{Ker} F^*_{n+1} \cong \operatorname{Ker} F^*_n \oplus E_n$$

where E_n is the trivial 1-dimensional vector bundle generated by the extra basis vector e_n. Thus

$$\operatorname{index} F \in K(X)$$

is well-defined independent of the choice of n.

Remarks. 1. Our definition of index F was dependent on a fixed orthonormal basis. It is a simple matter, which we leave to the reader, to show that the choice of basis is irrelevant.

2. Our introduction of K(X) here was motivated by the requirement of finding a natural "value group" for the index of Fredholm families. Historically the motivation for K(X) was somewhat different and arose from the work of Grothendieck in algebraic geometry. For this reason K(X) is referred to as the Grothendieck group of vector bundles over X.

Although our definition of the index as an element of K(X) may seem plausible, it is not clear at first how trivial or non-trivial it is. To show that it really is significant I will mention the following result.

THEOREM (2.3). *Let X be a compact space and let* $[X, \mathcal{F}]$ *denote the set of homotopy classes of continuous maps* $F : X \longrightarrow \mathcal{F}$. *Then* $F \longmapsto \operatorname{index} F$ *induces an isomorphism*

$$\operatorname{index} : [X, \mathcal{F}] \longrightarrow K(X) .$$

This theorem shows that our index is the only deformation invariant of a

Fredholm family. For example if X is a point the theorem asserts that the connected components of $\mathcal{F}$ correspond bijectively to the integers, the correspondence being given by the ordinary (integer) index. The proof of (2.3) is not deep. It can be reduced quite easily (see [1; Appendix (A6)]) to Kuiper's theorem that the group $\mathcal{a}^*$ of invertible operators on H is contractible. The proof of Kuiper's theorem, though ingenious, involves only elementary properties of Hilbert space.

§3. K-Theory.

Having introduced the group K(X) I shall now review briefly its elementary properties. In the first place we observe that one can form the tensor product $S \otimes T$ of two vector bundles and this induces a ring structure on K(X). Thus K(X) becomes a commutative ring with an identity element 1 -- corresponding to the trivial bundle $X \times \mathbb{C}^1$. When X is a point this is of course the usual ring structure of the integers $\mathbb{Z}$.

Next we investigate the behavior of K under change of parameter space. If $f : Y \longrightarrow X$ is a continuous map and if S is a vector bundle over X we obtain an induced vector bundle f^*S over Y, the fibre $f^*S(y)$ being $S(f(y))$. Passing to K we get a homomorphism of rings

$$f^* : K(X) \longrightarrow K(Y) \quad .$$

Thus K(X) is a contravariant functor of X. In this and other important respects K(X) closely resembles the cohomology ring H(X).

Finally we have the homotopy invariance of K(X) which asserts that $f^* : K(X) \longrightarrow K(Y)$ depends only on the homotopy class of the map $f : Y \longrightarrow X$. This follows from the fact that, in a continuous family of vector bundles over X, the isomorphism class is locally constant. For a simple proof of this see [1, (1.4.3)].

It is convenient to extend the definition of K(X) to locally compact spaces X by putting

$$K(X) = \mathrm{Ker}\{K(X^+) \xrightarrow{i^*} K(+)\}$$

where X^+ is the one-point compactification of X and $i : + \longrightarrow X^+$ is the inclusion of this "one-point." If X is not compact then $K(X)$ is a ring without identity: it is functorial for proper maps.

In addition to the ring structure in $K(X)$ it is convenient also to consider "external products." First of all, when X, Y are compact, we have a homomorphism

$$K(X) \otimes K(Y) \longrightarrow K(X \times Y)$$

obtained by assigning to a vector bundle E over X and a vector bundle F over Y the vector bundle $E \boxtimes F$ over $X \times Y$ whose fibre at a point (x, y) is $E_x \otimes F_y$. This extends at once to locally compact X, Y in view of the fact that we have an exact sequence

$$(3.1) \qquad 0 \longrightarrow K(X \times Y) \longrightarrow K(X^+ \times Y^+) \longrightarrow K(X^+) \oplus K(Y^+)$$

which identifies $K(X \times Y)$ with the subgroup of $K(X^+ \times Y^+)$ vanishing on the two "axes" X^+, Y^+. For a proof of (3.1) see [1, (2.48)].

If $x \in K(X)$, $y \in K(Y)$ the image of $x \otimes y$ in $K(X \times Y)$ is called their external product and written simply as xy.

As an important example of a locally compact space X we can take the Euclidean space $\mathbb{R}^n$, so that X^+ is the sphere S^n. Then we have

$$K(S^n) \cong K(\mathbb{R}^n) \oplus \mathbb{Z}$$

the $\mathbb{Z}$ summand being given by the dimension of a vector bundle. Thus $K(\mathbb{R}^n)$ is really the interesting part of $K(S^n)$. More generally for any X, (3.1) implies a decomposition

$$(3.2) \qquad K(S^n \times X) \cong K(\mathbb{R}^n \times X) \oplus K(X) \ .$$

The fundamental result of K-theory is the Bott periodicity theorem which asserts that the groups $K(\mathbb{R}^n)$ are periodic in n with period 2, so that

$$K(\mathbb{R}^{2m}) \cong K(\text{point}) = \mathbb{Z}$$
$$K(\mathbb{R}^{2m+1}) \cong K(\mathbb{R}^1) = 0$$

(the last equality depends on the fact that all vector bundles over the circle S^1 are trivial). This periodicity follows by induction from the more general result:

THEOREM 3.3. For any locally compact space X we have a natural isomorphism

$$K(\mathbb{R}^2 \times X) \cong K(X) \quad .$$

In §2 we explained the significant connection between the functor K and the space $\mathcal{F}$ of Fredholm operators. In view of this it is rather natural to raise the following two questions:

a) Can we use the index to prove the periodicity theorem (3.3)?

b) Can we use (3.3) to help study and compute indices?

The answer to both questions is affirmative. The first is remarkably simple and I will explain it in detail. The second is considerably deeper and I shall not comment on it further. What I have in mind will be found in [4]. A general discussion of the relationship between questions (a) and (b) is given in [3].

To carry out the program answering question (a) I will show, in the next section, how to define a homomorphism

$$\alpha : K(\mathbb{R}^2 \times X) \longrightarrow K(X)$$

using the index of certain simple Fredholm families. Once α has been constructed it is not difficult to show that it is an isomorphism, thus proving Theorem (3.3). This will be done in §5.

§4. The Wiener-Hopf operator.

I will recall here the discrete analogue of the famous Wiener-Hopf equation.[†]

Let $f(z)$ be a continuous complex-valued function on the unit circle $|z| = 1$. Let H be the Hilbert space of square-integrable functions on the circle $|z| = 1$ and let H_n be the closed subspace spanned by the functions z^k with $k \geq n$. In particular H_0 is the space of functions $u(z)$ with Fourier series of the form

$$u(z) = \sum_{k\geq 0} u_k z^k .$$

Let P denote the projection $H \longrightarrow H_0$ and consider the operator

$$F = Pf : H_0 \longrightarrow H_0$$

where f here stands for multiplication by f. Thus the Fourier coefficients $(Fu)_n$ $(n \geq 0)$ for $u \in H_0$ are given by

$$(4.1) \qquad (Fu)_n = \sum_{k\geq 0} f_{n-k} u_k \qquad (n \geq 0)$$

where f_m are the Fourier coefficients of f. The classical Wiener-Hopf equation is the integral counterpart of the discrete convolution equation (4.1), the summation being replaced by $\int_0^\infty \hat{f}(x-y)\hat{u}(y)dy$.

It is clear from the definition of F that

$$\|F\| \leq \sup|f(z)|$$

so that F depends continuously on f (for the norm topology of F and the sup norm topology of f). The basic result about these operators F is

PROPOSITION (4.2). *If $f(z)$ is nowhere zero (on $|z| = 1$) then $F = Pf : H_0 \longrightarrow H_0$ is a Fredholm operator.*

[†] For an exhaustive treatment of this topic see [5].

Proof. Let $\mathcal{C}$ denote the Banach algebra of all complex-valued functions f on the circle. Our construction $f \mapsto F$ defines a continuous linear map

$$\mathcal{C} \longrightarrow \mathcal{A}$$

where $\mathcal{A}$ is the Banach algebra of bounded operators on H_0. Passing to the quotient $\mathcal{B} = \mathcal{A}/\mathcal{K}$ by the ideal $\mathcal{K}$ of compact operators we obtain a continuous linear map $T : \mathcal{C} \longrightarrow \mathcal{B}$. Suppose now that $f, g \in \mathcal{C}$ and have <u>finite</u> Fourier series

$$f(z) = \sum_{-n}^{n} f_k z^k \qquad g(z) = \sum_{-m}^{m} g_k z^k$$

Let $\ell = fg$ and let F, G, L denote the corresponding elements of $\mathcal{A}$. It is then clear that

$$FG(z^k) = L(z^k) \quad \text{for} \quad k \geq m+n$$

i.e. the operators FG and L coincide on the subspace H_{m+n} of H_0. Thus FG - L has finite rank and so is compact. Passing to the quotient algebra $\mathcal{B}$ this means that

$$T(fg) = T(f)T(g) \quad .$$

Thus $T : \mathcal{C} \longrightarrow \mathcal{B}$ is a <u>homomorphism</u> on the subalgebra consisting of finite Fourier series. Since this subalgebra is dense in $\mathcal{C}$ and since T is continuous it follows that T itself is a homomorphism. Since T(1) = 1 this implies that T takes invertible elements into invertible elements. Thus, if f(z) is nowhere zero, T(f) is invertible in $\mathcal{B}$ and so, by (1.1), F is a Fredholm operator.

<u>Remark</u>. For the Fredholm operator F = Pf of this proposition an integer n for which $F_n = P_n f : H_0 \longrightarrow H_n$ is surjective can be found explicitly. It is enough to take a finite Fourier series

$$g(z) = \sum_{-n}^{n} g_k z^k$$

which approximates $f(z)^{-1}$ sufficiently so that

$$\sup|f(z)g(z) - 1| < 1 \ .$$

I shall omit the simple verification.

As a simple example consider the function $f(z) = z^m$. It is clear that for $F = Pf$ we then have

$$\text{index } F = -m \ .$$

Since the index is a locally constant function it follows that, for any continuous map

$$f : S^1 \longrightarrow \mathbb{C}^*$$

(where $\mathbb{C}^*$ denotes the non-zero complex numbers), we have

$$\text{index } F = -w(f) \tag{4.3}$$

where $w(f)$ is the "winding number" of f -- intuitively the number of times the path f goes round the origin in the complex plane. This simple observation is of fundamental importance for us. In fact the Bott periodicity theorem is, in a sense, a natural generalization of (4.3). We proceed now to introduce, by steps, the generalizations of the Wiener-Hopf operator which are required.

First we make an obvious extension, replacing the scalar-valued functions by vector-valued functions. The function f, which plays a multiplicative role, must be replaced by a matrix and the non-zero condition in (4.2) is replaced by the non-singularity of the matrix. Thus we start from a continuous map

$$f : S^1 \longrightarrow GL(N, \mathbb{C})$$

of the circle into the general linear group of $\mathbb{C}^N$. We take the Hilbert space of L^2-functions on S^1 with values in $\mathbb{C}^N$: if H is our original Hilbert space of scalar functions our new Hilbert space is $H \otimes \mathbb{C}^N$. As before we denote by P the projection

$$H \otimes \mathbb{C}^N \longrightarrow H_0 \otimes \mathbb{C}^N ,$$

and we define the operator

$$F = Pf : H_0 \otimes \mathbb{C}^N \longrightarrow H_0 \otimes \mathbb{C}^N$$

to be the composition of P and multiplication by the matrix function $f(z)$. The proof of (4.2) extends at once and shows that F is a Fredholm operator. The index of F is then a homotopy invariant of f. It determines in fact the element of the fundamental group of $GL(N, \mathbb{C})$ represented by f.

Next we generalize the situation by introducing a compact parameter space X. Thus we now consider a continuous map

$$f : S^1 \times X \longrightarrow GL(N, \mathbb{C}) \tag{4.4}$$

so that $f(z, x)$ is a non-singular matrix depending continuously on two variables z, x. For each $x \in X$ we therefore get a Fredholm operator $F(x)$, acting on the Hilbert space $H_0 \otimes \mathbb{C}^N$. Moreover $F(x)$ is a continuous function of x so that we have a Fredholm family

$$F : X \longrightarrow \mathcal{F} .$$

By the construction of §2 this family has an index in $K(X)$. Thus

$$f \longmapsto F \longmapsto \text{index } F$$

assigns to each continuous map f, as in (4.4), an element of $K(X)$. Moreover this element depends only on the homotopy class of f.

Our final generalization is to allow the vector space $\mathbb{C}^N$ to vary continuously with the parameter x. More precisely we fix an N-dimensional complex vector bundle V over X and we suppose given a function

$$(4.5) \qquad f(z, x) \in \operatorname{Aut} V(x)$$

(where $V(x)$ denotes the fibre of V over x) which is continuous in z, x. Since V is locally trivial our function f is locally of the type we had previously (with $V(x) = \mathbb{C}^N$ for all x), so that there is no problem in defining continuity. Our Hilbert space will now be $H_0 \otimes V(x)$ and so varies from point to point. The operator

$$F(x) : H_0 \otimes V(x) \longrightarrow H_0 \otimes V(x)$$

is defined as before and is a Fredholm operator depending continuously on x. Since our Hilbert space varies this is a somewhat more general kind of family than the Fredholm families of §2, but we can still define the index in $K(X)$ by essentially the same method. Since V is locally trivial we can still apply the purely local Lemma (2.1), replacing H_n by $H_n \otimes V(x)$, and obtaining locally an operator $F_n(x)$ with the properties described in (2.1). The compactness of X then leads to a fixed n for which $\operatorname{Ker} F_n$ is a vector bundle and $\operatorname{Ker} F_n^*$ is a trivial vector bundle. We now define

$$\operatorname{index} F = [\operatorname{Ker} F_n] - [\operatorname{Ker} F_n^*] \in K(X)$$

and prove as before that this is independent of n.

Finally therefore we have given a construction

$$(V, f) \longmapsto F \longmapsto \operatorname{index} F$$

which assigns to each pair (V, f) — consisting of a vector bundle V over X and an f as in (4.5) — an element of $K(X)$. Again this depends only on the homotopy class of f.

Now a pair (V, f) as above can be used to construct a vector bundle $E(V, f)$ over $S^2 \times X$. This is done as follows. We regard S^2 as the union of two closed hemi-spheres B^+, B^- meeting on the equator S^1. The vector bundle E is then constructed from the two vector bundles

$$\begin{array}{ccc} B^+ \times V & & B^- \times V \\ \downarrow & & \downarrow \\ B^+ \times X & & B^- \times X \end{array}$$

by identifying, over points $(z, x) \in S^1 \times X$,

$$(z, v) \in B^+ \times V(x) \text{ with } (z, f(z, x)v) \in B^- \times V(x) \quad .$$

When X is a point this is a well known construction for defining vector bundles over the sphere S^2. The parameter space X plays a quite harmless role.

It is not hard to show (see [1; p. 47]) that every vector bundle E over $S^2 \times X$ arises in this way from some pair (V, f). Given E we first take V to be the vector bundle over X induced from E by the inclusion map $x \mapsto (1, x)$ of X into $S^2 \times X$ (where 1 denotes the point $z = 1$ on $S^1 \subset S^2$). We then observe that, since $B^+ \times X$ retracts onto $\{1\} \times X$, the part E^+ of E over $B^+ \times X$ is isomorphic to $B^+ \times V$. Similarly $E^- \cong B^- \times V$. Then, over $S^1 \times X$, the identification of E^+ and E^- defines f. The map f obtained this way is normalized so that $f(1, x)$ is always the identity of $V(x)$, and its homotopy class is then uniquely determined by E. Hence our construction

$$E \mapsto (V, f) \longrightarrow F \mapsto \text{index } F$$

defines a map

$$\text{Vect}(S^2 \times X) \longrightarrow K(X) \quad .$$

This is clearly additive and so it extends uniquely to a homomorphism of groups

$$\alpha' : K(S^2 \times X) \longrightarrow K(X) \quad .$$

We recall now formula (3.2) which identifies $K(\mathbb{R}^2 \times X)$ with a subgroup of $K(S^2 \times X)$. Restricting α' to this subgroup we therefore obtain a homomorphism

$$\alpha : K(\mathbb{R}^2 \times X) \longrightarrow K(X) \quad .$$

Essentially therefore α is given by taking the index of a Wiener-Hopf family of Fredholm operators. From its definition it is clear that it has the following multiplicative property. Let Y be another compact space, then we have a commutative diagram

$$\text{(4.6)} \qquad \begin{array}{ccc} K(\mathbb{R}^2 \times X) \otimes K(Y) & \longrightarrow & K(\mathbb{R}^2 \times X \times Y) \\ \downarrow \alpha_X \otimes 1 & & \downarrow \alpha_{X \times Y} \\ K(X) \otimes K(Y) & \longrightarrow & K(X \times Y) \end{array}$$

where the horizontal arrows are given by the external product discussed in §3.

We can now easily extend α to locally compact spaces X by passing to the one-point compactification X^+ and using (3.1). The commutative diagram (4.6) continues to hold for X, Y locally compact — again by appealing to (3.1).

In the next section we shall show how to prove that α is an isomorphism, thus establishing the periodicity theorem (3.3).

§5. Proof of periodicity.

We begin by defining a basic element b in $K(\mathbb{R}^2)$. We take the 1-dimensional vector bundle E_m over S^2 defined, as in §4, by the function $f(z) = z^m$. We put

$$b = [E_{-1}] - [E_0] \in K(S^2) .$$

Since E_{-1} and E_0 both have dimension 1 it follows that b lies in the summand $K(\mathbb{R}^2)$ of $K(S^2)$. As we have already observed, if F_m is the Wiener-Hopf operator defined by the function z^m, we have

$$\text{index } F_m = -m .$$

Thus

$$\begin{aligned} \alpha(b) &= \text{index } F_{-1} - \text{index } F_0 \\ &= 1 . \end{aligned}$$

We now define, for any X, the homomorphism

$$\beta : K(X) \longrightarrow K(\mathbb{R}^2 \times X)$$

to be external multiplication by $b \in K(\mathbb{R}^2)$. Thus, for $x \in K(X)$, $\beta(x)$ is the image of $b \otimes x$ under the homomorphism

$$K(\mathbb{R}^2) \otimes K(X) \longrightarrow K(\mathbb{R}^2 \times X) .$$

With these preliminaries we can state a more precise version of the periodicity theorem.

THEOREM (5.1). *For any locally compact space* X, *the homomorphisms*

$$\beta : K(X) \longrightarrow K(\mathbb{R}^2 \times X)$$
$$\alpha : K(\mathbb{R}^2 \times X) \longrightarrow K(X)$$

are inverses of each other.

As we shall see the proof of (5.1) is now a simple consequence of the formal properties of α, β. First we apply the diagram (4.6) with X = point, Y = X. This gives, for any $x \in K(X)$,

$$\begin{aligned} \alpha\beta(x) &= \alpha(b)x \\ &= x \qquad \text{since } \alpha(b) = 1 \ . \end{aligned}$$

Thus α is a left inverse of β. To prove that it is also a right inverse we apply (4.6) with $Y = \mathbb{R}^2$. Thus we have the commutative diagram

$$\begin{array}{ccc} K(\mathbb{R}^2 \times X) \otimes K(\mathbb{R}^2) & \longrightarrow & K(\mathbb{R}^2 \times X \times \mathbb{R}^2) \\ \Big\downarrow \alpha_X & & \Big\downarrow \alpha_{X \times \mathbb{R}^2} \\ K(X) \otimes K(\mathbb{R}^2) & \longrightarrow & K(X \times \mathbb{R}^2) \end{array}$$

Hence for any element $u \in K(\mathbb{R}^2 \times X)$ we have

(5.2) $$\alpha(ub) = \alpha(u)b \in K(X \times \mathbb{R}^2) \ .$$

Consider now the map τ of $\mathbb{R}^2 \times X \times \mathbb{R}^2$ into itself which interchanges the two copies of $\mathbb{R}^2$. On $\mathbb{R}^4 = \mathbb{R}^2 \times \mathbb{R}^2$ this is given by the matrix

$$\begin{pmatrix} 0 & 0 & 1 & 0 \\ 0 & 0 & 0 & 1 \\ 1 & 0 & 0 & 0 \\ 0 & 1 & 0 & 0 \end{pmatrix} .$$

Since this has determinant +1 it lies in the identity component of $GL(4, \mathbb{R})$. Thus the map τ is homotopic to the identity (through homeomorphisms) and so τ^* is the identity of $K(\mathbb{R}^2 \times X \times \mathbb{R}^2)$. On the other hand we have

$$\tau^*(ub) = b\widetilde{u}$$

where $\widetilde{u} \in K(X \times \mathbb{R}^2)$ corresponds to $u \in K(\mathbb{R}^2 \times X)$ under the obvious identification. Thus

$$\alpha(ub) = \alpha(\tau^*(ub)) = \alpha(b\widetilde{u}) = \alpha\beta(\widetilde{u}) = \widetilde{u}$$

since α is a left inverse of β. Combined with (5.2) this gives

$$\widetilde{u} = \alpha(u)b \in K(X \times \mathbb{R}^2) \quad .$$

Switching back to $K(\mathbb{R}^2 \times X)$ this is equivalent to

$$u = b\alpha(u) = \beta\alpha(u) \in K(\mathbb{R}^2 \times X) \quad ,$$

proving that α is also a right inverse of β. This completes the proof of the theorem.

Concluding remarks.

I have given this proof of the periodicity theorem in such detail because I wanted to show how simple it really was. I hope I have demonstrated that the index of Fredholm families, which was used to construct the map α, plays a natural and fundamental role in K-theory.

This proof of periodicity has the advantage that it extends, with little effort, to various generalizations of the theorem. For full details on this I refer to [2].

REFERENCES

1. M. F. Atiyah, K-Theory (Benjamin, 1967).

2. ————————, "Bott periodicity and the index of elliptic operators," Quart. J. Math. Oxford, 1968 (to appear).

3. ————————, "Algebraic topology and elliptic operators," Comm. Pure Appl. Math. 20 (1967), 237-249.

4. ———————— and I. M. Singer, "The index of elliptic operators I," Ann. of Math., 1968 (to appear).

5. I. C. Gohberg and M. G. Krein, "Systems of integral equations on a half-line with kernels depending on the difference of arguments," Uspehi Mat. Nauk 13 (1958), 3-72. = Amer. Math. Soc. Translations (2) 14, 217-287.

Deformations of Riemann surfaces

by

Clifford J. Earle[(1)] and James Eells[(2)]

1. Introduction.

The transcendental theory of deformations of Riemann surfaces has been greatly enriched in the past decade, primarily through the contributions of L. V. Ahlfors and L. Bers. Their point of departure lay in fundamental ideas of O. Teichmüller - who defined a natural covering of the Riemann space of moduli (for a closed surface of genus g) by a Euclidean cell, which is today called the Teichmüller space $\mathfrak{T}(g)$ for genus g. Much effort has been spent in determining the structure of $\mathfrak{T}(g)$, and its relationship to spaces of holomorphic quadratic differentials on Riemann surfaces.

In the present article we describe briefly one approach to that subject. Our main step (which is established in detail in [28]) is the construction of a certain principal fibre bundle whose base space is $\mathfrak{T}(g)$, and whose structural group depends only on g. Many interesting questions related to deformation theory can be formulated conveniently in terms of fibre bundles associated with our principal bundle (see §§3A, 5B). For instance, the infinitesimal variation theory (for surfaces) of Kodaira-Spencer [42] appears as the tangent vector bundle sequence associated with the differential of our bundle map. That permits us to identify the tangent space to $\mathfrak{T}(g)$ at a given point with

(1) Research partially supported by NSF Grant GP-6145.
(2) Research partially supported by NSF Grant GP-4216.

the appropriate space of holomorphic quadratic differentials.

Two elliptic operators play fundamental roles in our theory; they are discussed in §4. The first is Beltrami's equation (a first order linear elliptic differential equation, which in our treatment has smooth coefficients); that equation lies at the heart of quasiconformal mapping theory. The second is the tension equation (a second order quasi-linear operator derived as the variational equation of the Dirichlet-Douglas integral). A primary aim of our presentation is to show how the various analytical properties of these operators reflect differential topological aspects of our fibre bundle.

In §5 we describe a few applications of Teichmüller theory: to differential topology, complex function theory, and to differential geometry. These are chosen to illustrate the wide range of applicability of the theory; they are by no means exhaustive. For instance, we have not discussed the simultaneous uniformization of closed surfaces of genus g [16,18].

With the help of quasiconformal mappings Teichmüller theory can be extended to open Riemann surfaces. We have indicated the method briefly in §§4C, D. The reader can learn about the general theory from Ahlfors [4,5] and Bers [17,18]. In addition to those, the reader should consult Bers [13,14], Earle-Eells [28], Rauch [54], and Teichmüller [59] for more detailed accounts of the theory for closed surfaces and for guides to the literature.

2. Teichmüller theory for closed surfaces.

(A) Let S be a closed oriented 2-dimensional manifold. The homeomorphism type of S is of course determined by its genus g = the number of handles of S. It is well known that S admits a smooth structure (of class C^∞) which is essentially unique.

If T_xS is the oriented tangent space of S at the point $x \in S$, then a complex structure on T_xS is an $\mathbb{R}$-automorphism $J_x : T_xS \to T_xS$ which satisfies $J_x^2 = -I$ and which is compatible with the given orientation of T_xS. The effect of J_x is to make T_xS a 1-dimensional complex vector space, with $iv = J_x(v)$. A _complex structure on_ S is a smooth (= C^∞) assignment $x \to J_x$. (Such an assignment is usually called an almost complex structure, but on S the almost complex and the complex structures coincide.) Every S admits complex structures [58]. They can be partitioned into equivalence classes, which form families of complex dimension

$$\begin{aligned} 0 \quad &\text{if} \quad g = 0, \\ 1 \quad &\text{if} \quad g = 1, \\ 3g-3 \quad &\text{if} \quad g \geq 2. \end{aligned}$$

To make that statement precise is the moduli problem; it is our purpose to describe one approach to that problem.

(B) Assume that S has genus $g \geq 2$. _Let_ $\mathfrak{M}(g)$ _denote the totality of complex structures on_ S, _with its_ C^∞-_topology_ (i.e., we think of the complex structures as smooth tensor fields on S and define convergence as uniform convergence of each iterated differential).

There are alternative descriptions of $\mathcal{M}(g)$ which will be convenient at times: We can view $\mathcal{M}(g)$ as the space (with C^∞-topology) of conformal equivalence classes of smooth Riemannian structures on S; or as the space (with C^∞-topology) of Riemannian structures on S with constant curvature = -4.

Whatever else, $\mathcal{M}(g)$ is a contractible complex analytic manifold (therefore an absolute retract) modeled on a separable Fréchet space [28, §5].

Let $\mathcal{D}(g)$ be the group of all orientation-preserving C^∞-diffeomorphisms of S, with its C^∞-topology. $\mathcal{D}(g)$ is a topological group and a metrizable absolute neighborhood retract [28]. We define $\mathcal{D}_0(g)$ as the closed normal subgroup consisting of the diffeomorphisms which are homotopic to the identity. That definition has a temporary character, for in fact $\mathcal{D}_0(g)$ is the arc component of the identity in $\mathcal{D}(g)$ (see §2C below).

Viewing $\mathcal{M}(g)$ as a space of tensor fields, we have a natural action of $\mathcal{D}(g)$ on $\mathcal{M}(g)$

(2.1) $$\mathcal{M}(g) \times \mathcal{D}(g) \to \mathcal{M}(g),$$

which we write $(J,\varphi) \to J\cdot\varphi$. By general principles, that action is continuous.

(C) Theorem [26,28]. Assume that genus(S) = $g \geq 2$. Then

(a) $\mathcal{D}(g)$ acts effectively and properly on $\mathcal{M}(g)$.

(b) If

(2.2) $$\Phi: \mathcal{M}(g) \to \mathcal{M}(g)/\mathcal{D}_0(g) = \mathcal{T}(g)$$

is the quotient map (with quotient topology on $\mathcal{T}(g)$), then (2.2) is a universal principal $\mathcal{D}_0(g)$-fibre bundle.

To say the action is <u>effective</u> means that if $\varphi \in \mathcal{D}(g)$ and $J\cdot\varphi = J$ for all $J \in \mathcal{M}(g)$, then $\varphi = I$. The action is <u>proper</u> if $\Theta(J,\varphi) = (J,J\cdot\varphi)$ defines a proper map $\Theta\colon \mathcal{M}(g) \times \mathcal{D}(g) \to \mathcal{M}(g) \times \mathcal{M}$

(2.2) is called a <u>principal</u> $\mathcal{D}_0(g)$-fibre bundle if $\mathcal{D}_0(g)$ acts freely, properly, and Φ has local sections. A principal bundle is <u>universal</u> if its total space (in this case $\mathcal{M}(g)$) is contractible. $\mathcal{D}_0(g)$ acts freely because a closed Riemannian manifold with negative curvature has at most one isometry in any homotopy class. The proper action is a consequence of (a) because $\mathcal{D}_0(g)$ is a closed subgroup of $\mathcal{D}(g)$. Φ is said to <u>have local sections</u> if for each $\tau \in \mathcal{T}(g)$ there exist a neighborhood V and a continuous map $s\colon V \to \mathcal{M}(g)$ with $\Phi \circ s = I$. The existence of local sections follows from the results of §4. In §4E we exhibit a globally defined section. That provides one proof of the next

<u>Theorem [28]</u>. There is a continuous section of the bundle (2.2). In particular, there is a homeomorphism

$$\mathcal{M}(g) \to \mathcal{T}(g) \times \mathcal{D}_0(g). \tag{2.3}$$

An alternative proof that there is a section can be obtained from Teichmüller's Theorem 2D. The homeomorphism (2.3) follows because a principal bundle admits a section when and only when it is a trivial bundle.

<u>Corollary [28]</u>. Assume that genus(S) = $g \geq 2$. Then $\mathcal{D}_0(g)$ is contractible.

In particular, $\mathcal{D}_0(g)$ is arcwise connected. The corollary is an immediate consequence of the homeomorphism (2.3). We shall give a more complete discussion of the groups $\mathcal{D}(g)$ in §5A.

(D) The quotient space $\mathfrak{T}(g)$ in (2.2) is called <u>Teichmüller's space for genus</u> g. A great deal is known about its structure.

<u>Theorem</u>. $\mathfrak{T}(g)$ is a topological manifold, homeomorphic to Euclidean space of dimension 6g-6. It has a complex structure such that $\Phi: \mathfrak{M}(g) \to \mathfrak{T}(g)$ is holomorphic. The complex manifold $\mathfrak{T}(g)$ is holomorphically equivalent to a bounded domain of holomorphy in $\mathbb{C}^{3g-3}$; in particular, $\mathfrak{T}(g)$ is a Stein manifold.

The first assertion is known as Teichmüller's theorem [14,59], although earlier versions, based on the theory of Fuchsian groups, were given by Fricke-Klein, Vorlesungen über die Theorie der automorphen Funktionen 1, Leipzig, 1926 (Zweiter Abschnitt), and in the unpublished manuscript of Nielsen and Fenchel. The existence of a natural complex structure was first demonstrated by Ahlfors [1], then by Bers [13], using the complex structure of $\mathfrak{M}(g)$. Partial results had been obtained by Rauch [50]. Bers [15] imbedded $\mathfrak{T}(g)$ in $\mathbb{C}^{3g-3}$ as a bounded domain. Bers and Ehrenpreis [19] showed that the image is a domain of holomorphy.

Further, $\mathfrak{T}(g)$ is taut, in the sense that for any finite dimensional complex manifold X, the family of holomorphic maps from X to $\mathfrak{T}(g)$ is normal (that implies domain of holomorphy [61]). Presumably $\mathfrak{T}(g)$ is not homogeneous for $g > 1$.

(E) The <u>Riemann space for genus</u> g is the quotient space $\mathcal{R}(g) = \mathfrak{M}(g)/\mathcal{D}(g)$, produced by the action (2.1). $\mathcal{R}(g)$ may be viewed as the space of (conformal) equivalence classes of Riemann surfaces of genus g, the so-called space of moduli. Obviously

$$\mathcal{R}(g) = \mathcal{M}(g)/\mathcal{D}(g) = \mathcal{T}(g)/\Gamma(g),$$

where $\Gamma(g) = \mathcal{D}(g)/\mathcal{D}_o(g)$ is the Teichmüller modular group for genus g. Since $\mathcal{D}_o(g)$ is open and closed in $\mathcal{D}(g)$, $\Gamma(g)$ is a discrete topological group.

Proposition [43]. The modular group $\Gamma(g)$ acts properly discontinuously on $\mathcal{T}(g)$ as a group of holomorphic transformations

Because $\mathcal{T}(g)$ is a Stein manifold, general principles [21] imply that $\mathcal{R}(g)$ is a normal analytic space. Baily [12] has shown that $\mathcal{R}(g)$ is quasi-projective (i.e., holomorphically equivalent to a space of the form A-B, where A and B are projective complex algebraic varieties).

$\Gamma(g)$ does not act freely on $\mathcal{T}(g)$; in fact, every finite subgroup $\Gamma \subset \Gamma(g)$ has fixed points [43]! The fixed point locus of Γ is a closed complex submanifold of $\mathcal{T}(g)$, homeomorphic to a cell [43,54]. With few exceptions the fixed points lie over non-manifold points of $\mathcal{R}(g)$ [53].

The modular group $\Gamma(g)$ is isomorphic to the outer automorphism group of the fundamental group $\pi_1(S)$; i.e.,

$$\Gamma(g) \approx \text{Aut}(\pi_1(S))/\text{Inner}(\pi_1(S)),$$

where $\text{Aut}(\pi_1(S))$ is the group of automorphisms of the fundamental group of S and Inner(π_1(S)) is the normal subgroup of inner automorphisms [10,11]. Various convenient sets of generators can be given for $\Gamma(g)$; see [45], [47], [48]. In particular, it is known [48] that $\Gamma(g)$ can be generated by four elements. Also [48], that $\Gamma(g)$ modulo its commutator subgroup is a finite cyclic group whose order divides 10.

Little is known about the homology or homotopy groups of $\mathcal{R}(g)$. The fundamental group $\pi_1(\mathcal{R}(g))$ is isomorphic to $\Gamma(g)$ modulo its torsion subgroup [9]. $H_1(\mathcal{R}(g))$ is an abelian homomorphic image of $\Gamma(g)$, hence finite cyclic. $H_i(\mathcal{R}(g)) = 0$ for $i > 3g-3$ by general Stein theory.

(F) Every subgroup $\mathcal{D}'$ of $\mathcal{D}(g)$ which contains $\mathcal{D}_o(g)$ determines a complex analytic space $\mathcal{M}(g)/\mathcal{D}'$ lying between $\mathfrak{J}(g)$ and $\mathcal{R}(g)$. Two of these spaces are of particular interest. The <u>Torelli space</u> $\tau(g) = \mathcal{M}(g)/\mathcal{D}_1(g)$, where $\mathcal{D}_1(g) = \{\varphi \in \mathcal{D}(g)$: φ induces the identity map on the integral homology groups of S$\}$. A theorem of Hurwitz [40] insures that $\mathcal{D}_1(g)$ acts freely on $\mathcal{M}(g)$. From §2C,D,E it follows that

$$\Phi_1: \mathcal{M}(g) \to \mathcal{M}(g)/\mathcal{D}_1(g) = \tau(g)$$

<u>is a universal principal $\mathcal{D}_1(g)$ - fibre bundle</u>, and the natural map π: $\mathfrak{J}(g) \to \tau(g)$ is a <u>holomorphic covering map</u>. In particular, $\tau(g)$ is a complex analytic manifold [21]. Since $\mathfrak{J}(g)$ is contractible, the fundamental group $\pi_1(\tau(g)) = \mathcal{D}_1(g)/\mathcal{D}_o(g)$ completely determines the homotopy type of $\tau(g)$.

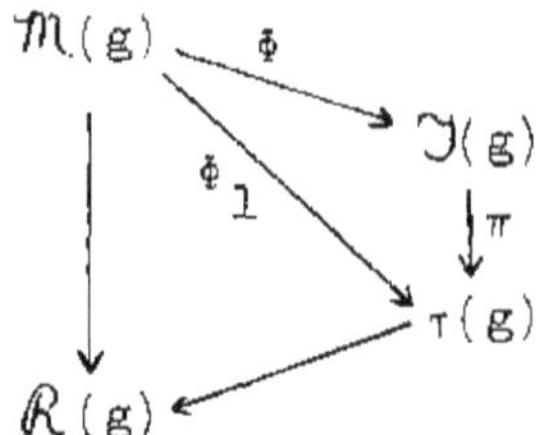

The <u>Siegel space</u> $\sigma(g) = \mathcal{M}(g)/\mathcal{D}_2(g)$, where $\mathcal{D}_2(g)$ is the group generated by $\mathcal{D}_1(g)$ and an involution of S which carries some canonical homology basis to its negative. The action of $\mathcal{D}_2(g)$ on $\mathcal{M}(g)$ is not free; the points with non-trivial stabilizer correspond to the hyperelliptic surfaces. The natural map from $\tau(g)$ onto $\sigma(g)$ is generically 2-1 but is 1-1 on the hyperelliptic locus.

Regarding a point of $\tau(g)$ as a Riemann surface with a canonical homology basis, we can map each point to its Riemann period matrix [12]. That determines a holomorphic map [18] from $\tau(g)$ into the Siegel half-plane H(g), the space of symmetric matrices in $gl(\mathbf{C}^g)$ with positive definite imaginary part. (The requirement that this map be holomorphic uniquely determines the complex structures of $\tau(g)$ and $\mathcal{J}(g)$ [1].) According to Torelli's theorem [8] there is an identification of $\sigma(g)$ and the image of $\tau(g)$ in H(g). Under that correspondence the action of $\mathcal{D}(g)/\mathcal{D}_2(g)$ on $\sigma(g)$ is the restriction of the familiar action of the symplectic modular group $Sp(\mathbf{Z}^g)$ on H(g). We therefore obtain a 1-1 holomorphic map from the Riemann space $\mathcal{R}(g)$ into the normal complex space $H(g)/Sp(\mathbf{Z}^g)$. According to Baily [12] the image of $\mathcal{R}(g)$ is a normal complex subspace of $H(g)/Sp(\mathbf{Z}^g)$, holomorphically equivalent to $\mathcal{R}(g)$.

3. Infinitesimal theory for closed surfaces.

(A) First of all, we consider the tangent vector bundle of $\mathcal{J}(g)$.

Each complex structure $J \in \mathcal{M}(g)$ determines a (p,q)-decomposition of the complexified tangent bundle $CT(S) = T^{1,0}(S) \oplus T^{0,1}(S)$. If A^p denotes the vector space of smooth differential forms on S of type (0,p) with values in $T^{1,0}(S)$, the exterior differential operator $\bar{\partial}_J$ (determined by the complex structure J) maps $A^p \to A^{p+1}$ with $\bar{\partial}_J^2 = 0$. Then $A^1 = \operatorname{Ker} \bar{\partial}_J$, and A^0 can be identified with the vector space of smooth vector fields on S. The quotient space $A^1/\bar{\partial}_J A^0 = H^1(S,\Theta_J)$, where Θ_J is the sheaf of germs of smooth sections of $T^{1,0} \otimes T^{*0,1}$. $H^1(S,\Theta_J)$ is the cohomology space used by Kodaira-Spencer [42] to study the infinitesimal variations of J.

The differential

$$T_J\Phi\colon T_J \mathcal{M}(g) = A^1 \to T_{\Phi J} \mathcal{J}(g)$$

has kernel $\bar{\partial}_J A^0$, so that we have an identification $T_{\Phi J} \mathcal{J}(g) = H^1(S,\Theta_J)$.

Serre's duality theorem asserts that $H^1(S,\Theta_J)$ has the space $\mathcal{B}_2(S,J)$ of J-holomorphic quadratic differentials on S as conjugate space. The Riemann-Roch theorem tells us that

$$\dim_{\mathbb{C}} \mathcal{B}_2(S,J) = 3g-3.$$

<u>Thus the space</u> $\mathcal{B}_2(g) = \cup\{\mathcal{B}_2(S,J)\colon J \in \mathcal{M}(g)\}/\mathcal{D}_0(g)$ <u>has</u>

natural identification as a vector bundle; in fact, the tangent bundle T $\mathfrak{J}(g)$.

(B) The tangent space $T_J \mathcal{M}(g) = A^1$ has a natural inner product relative to the Poincaré metric determined by J. Represent S as $\Gamma \backslash U$ where $U = \{z \in \mathbb{C}: \operatorname{Im}(z) > 0\}$; Γ is a Fuchsian group of Möbius transformations operating on U, and the identification $S \to \Gamma \backslash U$ is J-holomorphic. Then the inner product on A^1 takes the form

$$(3.1) \qquad \langle \mu, \nu \rangle_J = \int_{\Gamma \backslash U} \mu \bar{\nu} \, dm,$$

where $dm(z) = \rho(z)^2 \, dxdy$ is the Poincaré area element. If we map $\mathcal{B}_2(S,J)$ into A^1 by $\varphi \to \bar{\varphi}\rho^{-2}$, we obtain the orthogonal decomposition

$$T_J \mathcal{M}(g) = \operatorname{Ker} T_J \Phi \oplus \mathcal{B}_2(S,J).$$

The restriction of the inner product (3.1) to $\mathcal{B}_2(S,J)$ is the Petersson inner product of holomorphic quadratic differentials

$$(3.2) \qquad \langle \varphi, \psi \rangle_J = \int_{\Gamma \backslash U} \varphi \bar{\psi} \rho^{-4} dm = \int_S \varphi * \bar{\psi}.$$

By the above identification of $\mathcal{B}_2(S,J)$ with $T_{\Phi J} \mathfrak{J}(g)$, the inner product (3.2) yields a natural Kählerian structure on $\mathfrak{J}(g)$. It has been shown by Ahlfors [3] that (3.2) has negative Ricci curvature and negative holomorphic sectional curvature. The modular group $\Gamma(g)$, because of the naturality, acts as a group of isometries. It is not known whether the metric on $\mathcal{J}(g)$ associated with (3.2) is complete.

(C) We next identify the space $H^1(S,\Theta_J)$ of infinitesimal variations of complex structure with deformations of Fuchsian groups.

We mark the surface S by choosing a base point $x_o \in S$ and a canonical system of loops $a_1,\dots,a_g$; $b_1,\dots,b_g$ generating the fundamental group $\pi_1(S,x_o)$. If $U = \{z \in \mathbb{C}: \mathrm{Im}(z) > 0\}$, then for each $J \in \mathcal{M}(g)$ there exist a unique J-holomorphic covering map $\pi: U \to S$ and base point $z_o \in \pi^{-1}(x_o)$ such that the cover transformations A_1 and B_1 determined by a_1 and b_1 have fixed points at 0, 1, and ∞; A_1 fixing 0 and ∞, B_1 having attractive fixed point 1. The group Γ of cover transformations has generators $A_1,\dots,B_g$ determined by the marking $a_1,\dots,b_g$.

Let $G = SL(\mathbb{R}^2)/\text{Center}$ be the group of all holomorphic automorphisms of U. Then <u>the map</u> $P: \mathcal{M}(g) \to G^{2g}$ <u>defined by</u> $P(J) = (A_1,B_1,\dots,A_g,B_g)$ <u>is constant on</u> $\mathcal{D}_o(g)$-<u>orbits</u> and therefore determines a map $\overline{P}: \mathcal{J}(g) \to G^{2g}$. $\overline{P}$ is in fact a diffeomorphism from $\mathcal{J}(g)$ onto an open subset of the non-singular locus of an algebraic variety in G^{2g}.

If $\mathfrak{g}$ denotes the Lie algebra of G we define $H^1(\Gamma,\mathfrak{g})$ as the 1-dimensional cohomology space of Γ relative to the adjoint representation of $\mathfrak{g}$ on G. That space measures the infinitesimal deformations of Γ in G.

<u>Theorem [28]</u>. The differential $T_JP: A^1 \to H^1(\Gamma,\mathfrak{g})$ of $P: \mathcal{M}(g) \to G^{2g}$ at J induces an isomorphism $H^1(S,\Theta_J) \to H^1(\Gamma,\mathfrak{g})$.

(D) Let Π_n be the real vector space of polynomial functions on U of degree $\leq$ n having real coefficients ($n \geq 0$). The <u>Eichler action</u> of G on Π_{2q-2}, $q \geq 2$, is defined by

$$P^A(z) = P(Az)\ A'(z)^{1-q},\quad A \in G,\ z \in U.$$

It is well known that the Lie algebra $\mathfrak{g}$ of G is isomorphic to Π_2 through the map

$$X \to pX(z) = \lim_{t\to 0} \frac{\exp\ tX(z)-z}{t},\quad z \in U.$$

<u>Under the map the adjoint action of G on $\mathfrak{g}$ and the Eichler action of G on Π_2 coincide.</u> Thus, Theorem 3C gives an isomorphism between $H^1(S,\Theta_J)$ and the Eichler cohomology space $H^1(\Gamma,\Pi_2)$.

For any integer $q \geq 2$ let A_q^p be the vector space of smooth differential forms on S of type (0,p) with values in the vector bundle $\odot^{q-1} T^{1,0}(S)$. Again we have the coboundary operator $\bar{\partial}_J\colon A_q^o \to A_q^1$. Let Θ_q be the sheaf of germs of smooth sections of $\odot^{q-1} T^{1,0}(S) \otimes T^{*0,1}(S)$. Then $H^1(X,\Theta_q) = A_q^1/A_q^o$. By Serre duality, $H^1(S,\Theta_q)$ is conjugate to $H^o(S, \odot^q T^{*1,0}) = \mathcal{B}_q(S,J)$, the vector space of J-holomorphic q-adic differentials on S.

Let $H^1(\Gamma,\Pi_{2q-2})$ be the cohomology group determined by the Eichler action of Γ on Π_{2q-2}. Theorem 3C has this generalization [17]: <u>There is a natural linear map</u> $P_q\colon A_q^1 \to H^1(\Gamma,\Pi_{2q-2})$ <u>which induces an isomorphism</u> $H^1(S,\Theta_q) \to H^1(\Gamma,\Pi_{2q-2})$.

4. Analytical techniques.

(A) Let D be a subregion of the complex plane $\mathbb{C}$. Every smooth Riemannian metric on D can be written uniquely in the form

$$ds^2 = \lambda(z)\,|dz + \mu(z)\,d\bar{z}|^2 \tag{4.1}$$

where $\lambda(z)$ is positive and $|\mu(z)| < 1$. The conformal equivalence class of that metric is fully determined by the function μ. Thus we obtain an identification between the spaces $\mathcal{M}(D)$ and $C^\infty(D,\Delta)$, where $\Delta = \{z \in \mathbb{C}\colon |z| < 1\}$.

If D_μ denotes D with the complex structure determined by $\mu \in C^\infty(D,\Delta)$, then $w\colon D_\mu \to \mathbb{C}_0$ is holomorphic if and only if it satisfies the Beltrami equation

$$w_{\bar{z}} = \mu(z) w_z, \tag{4.2}$$

where $w_{\bar{z}} = (\partial/\partial\bar{z})w = \frac{1}{2}\left(\frac{\partial}{\partial x} + i\,\frac{\partial}{\partial y}\right)w$, and $w_z = (\partial/\partial z)w = \frac{1}{2}\left(\frac{\partial}{\partial x} - i\,\frac{\partial}{\partial y}\right)w$. The equation (4.2) is elliptic because $|\mu(z)| < 1$ in D. It is uniformly elliptic in D provided that $\|\mu\| = \sup\{|\mu(z)|\colon z \in D\} < 1$. Under that condition the following theorems hold.

1) The equation (4.2) has a solution which is a diffeomorphism of D onto a region in $\mathbb{C}$. If $D = U = \{z \in \mathbb{C}\colon \mathrm{Im}(z) > 0\}$, there is a unique solution w_μ of (4.2) which is a homeomorphism of $\bar{U}$ onto itself leaving 0, 1, ∞ fixed. [6]

2) <u>For each</u> $k < 1$ <u>the map</u> $\mu \to w_\mu$ <u>is a homeomorphism of</u> $\mathcal{M}_k(U)$, <u>the set of</u> $\mu \in \mathcal{M}(U)$ <u>with</u> $\sup\{|\mu(z)|: z \in U\} \leq k < 1$, <u>onto its image in</u> $C^\infty(U,\mathbb{C})$. [28,29]

(B) According to the uniformization theorem, for any closed surface S with genus(S) = g $\geq$ 2 there is a covering map $\pi: U \to S$ with a cover group Γ consisting of Möbius transformations. The map π induces an identification of $\mathcal{M}(g)$ with the Γ-invariant $\mu \in \mathcal{M}(U)$. These are precisely the functions $\mu \in C^\infty(U,\Delta)$ which satisfy

$$(4.3) \qquad \mu(\gamma z)\overline{\gamma'(z)}/\gamma'(z) = \mu(z), \; \forall\, z \in U, \; \gamma \in \Gamma.$$

We call such functions <u>smooth Beltrami differentials</u> and denote their totality by $\mathcal{M}(\Gamma)$. Let $A^1(\Gamma)$ be the closed linear subspace of $C^\infty(U,\mathbb{C})$ consisting of all functions $\mu \in C^\infty(U,\mathbb{C})$ which satisfy (4.3). Since Γ has a compact fundamental domain in U, $\mathcal{M}(\Gamma)$ <u>is a convex open set in</u> $A^1(\Gamma)$. Further, <u>the map</u> $\mu \to w_\mu$ <u>is a homeomorphism from</u> $\mathcal{M}(\Gamma)$ <u>onto its image in</u> $C^\infty(U,\mathbb{C})$.

(C) The solution theory of the Beltrami equation, with the help of the above identification of $\mathcal{M}(g)$ with $\mathcal{M}(\Gamma)$, provides a short path to the theorems of §2C,D,E [28]. For some purposes it is useful to replace the Fréchet space $A^1(\Gamma)$ by the Banach space $L^\infty(\Gamma)$ of all bounded measurable functions on U which satisfy (4.3), with the usual supremum norm. The open unit ball $M(\Gamma)$ is the space of <u>bounded measurable Beltrami differentials</u>.

For $\mu \in M(\Gamma)$, the equation (4.2) has a unique (distribution) solution w_μ which is a homeomorphism of $\overline{U}$ onto itself fixing 0, 1, ∞. w_μ is called a <u>quasiconformal</u> map. Its distribution derivatives w_z, $w_{\bar{z}}$ are locally L^p functions, for some $p > 2$ [6].

The chief advantage of quasiconformal maps over smooth maps is that <u>if</u> $k < 1$, <u>the set of all</u> w_μ <u>with</u> $\|\mu\| \leq k < 1$ <u>is compact with respect to uniform convergence on compact sets in</u> $\overline{U}$. Moreover, the use of quasiconformal maps permits the construction of Teichmüller spaces for arbitrary Fuchsian groups Γ (or, almost equivalently, for arbitrary surfaces $\Gamma\backslash U$).

Briefly, the totality of quasiconformal maps of $S = \Gamma\backslash U$ is a group $Q(\Gamma)$, and those which are homotopic to the identity form a normal subgroup $Q_0(\Gamma)$. $Q(\Gamma)$ acts on $\mathfrak{M}(\Gamma)$ as a group of holomorphic automorphisms. The quotient space $\mathfrak{T}(\Gamma) = M(\Gamma)/Q_0(\Gamma)$ is <u>Teichmüller's space for the Fuchsian group</u> Γ. If Γ is the cover group of a closed surface of genus g, then $\mathfrak{T}(\Gamma) = \mathfrak{T}(g)$. As an analogue of Theorem 2C, we have [25,27]: <u>The quotient map</u>

$$(4.4) \qquad \Phi: M(\Gamma) \to \mathfrak{T}(\Gamma) = M(\Gamma)/Q_0(\Gamma)$$

<u>is a locally trivial fibre space.</u> But $Q_0(\Gamma)$ with the topology determined by Φ is <u>not a topological group</u>, so (4.4) does not define a $Q_0(\Gamma)$-fibre bundle.

$\mathfrak{T}(\Gamma)$ has a natural complex structure such that Φ is holomorphic [18]. The modular group $Q(\Gamma)/Q_0(\Gamma)$ operates as a group of holomorphic automorphisms [18], but does not always act properly discontinuously on $\mathfrak{T}(\Gamma)$.

(D) The space $M(\Gamma)$ carries a natural Finsler structure [25] which is invariant under all its holomorphic automorphisms. The induced metric is given by

$$d(\mu,\nu) = \text{ess sup}\{\sigma(\mu(z),\nu(z)) : z \in U\}$$

where σ is the Poincaré distance in Δ. The quotient metric on $\mathfrak{J}(\Gamma)$ produced by the map Φ is Teichmüller's metric. It is a complete metric, relative to which the modular group is a group of isometries.

Kravetz has shown that $\mathfrak{J}(g)$ with Teichmüller's metric is a straight space of negative curvature in the sense of Busemann [20]. The fact that each finite subgroup of the modular group $\Gamma(g)$ has fixed points in $\mathfrak{J}(g)$ is an important consequence of the negative curvature [43].

(E) Let us now interpret the elements of $\mathcal{M}(g)$ as smooth Riemannian metrics of constant curvature -4 on the closed surface S. Given μ, $\nu \in \mathcal{M}(g)$ and a smooth map $f: S \to S$ we form its energy (or Dirichlet-Douglas integral)

$$E(f) = \frac{1}{2} \int_S \sigma^2(f(z)) \, [|f_z|^2 + |f_{\bar{z}}|^2] dxdy.$$

Here $z = x + iy$ is an isothermal parameter relative to μ, $ds = \rho(z)|dz|$, and ν is given in isothermal parameters by $ds = \sigma(w)|dw|$.

The tension field of f (or Euler-Lagrange operator associated with E) is

$$\tau(f) = \frac{4}{\rho^2}\,[w_{z\bar{z}} + 2\,\frac{\sigma_w}{\sigma}\,w_z\,w_{\bar{z}}].$$

This is a second order elliptic quasi-linear system. The smooth extremals of E are called harmonic maps.

The second variation of E is

$$\frac{d^2E(f_t)}{dt^2} = \int_S g^{ij} \langle \frac{D}{\partial x^i}(\frac{\partial f_t}{\partial t}), \frac{D}{\partial x^j}(\frac{\partial f_t}{\partial t})\rangle *1$$

$$- \int_S g^{ij} \langle \mathcal{R}(\frac{\partial f_t}{\partial x^i}, \frac{\partial f_t}{\partial t})\frac{\partial f_t}{\partial t}, \frac{\partial f_t}{\partial x^j}\rangle *1 - \int_S \langle \frac{D}{\partial t}(\frac{\partial f_t}{\partial t}), \tau(f_t)\rangle *1,$$

where we now write $ds^2 = g_{ij}dx^i dx^j$, and let $*1$ denote its volume element, and $D/\partial x^i$ a covariant derivative.

In every homotopy class of maps $f\colon X \to X$ there is a unique harmonic map $f(\mu,\nu)$. The existence was proved in [30,56]; the uniqueness in [7,38].

That harmonic map $f(\mu,\nu)$ is bijective [56], and in fact is a diffeomorphism [44,39].

Thus for any fixed map $\mu \in \mathcal{M}(g)$ we have a map $\mathcal{M}(g) \to \mathcal{D}_0(g)$ defined by $\nu \to f(\mu,\nu)$; Sampson [55] has shown that that map is continuous. These ideas can be used to prove that the map $\Phi\colon \mathcal{M}(g) \to \mathcal{T}(g)$ in §2C has a global section [28], without appeal to Teichmüller's theorem.

5. *Various applications.*

(A) The theorems in §2C have immediate topological implications for $g \geq 2$, using the homotopy exact sequence of a fibration. Furthermore, it is not difficult (with the conformal interpretation of $\mathfrak{M}(g)$ in place of the complex structure interpretation) to modify the constructions to include the cases $g = 1, 0$ as well as the nonorientable surfaces. We extend our notation, letting $\mathcal{D}(S) = \mathcal{D}(g)$ denote the topological group of all diffeomorphisms of S, if S is nonorientable.

First of all, we have the following result, which is the C^∞-analogue of a topological theorem of R. Baer [10,11]:

$\mathcal{D}_o(S)$ *is the arc component in* $\mathcal{D}(S)$ *of the neutral element.*

Next, there is the homotopy description of $\mathcal{D}_o(S)$ [26,28]:

1) *If* S *is the sphere or projective plane, then* $\mathcal{D}(S) = \mathcal{D}_o(S$ *has the rotation group* SO(3) *as strong deformation retract.*

2) *If* S *is the torus, then* $\mathcal{D}_o(S)$ *has* $S = SO(2) \times SO(2)$ *as strong deformation retract.*

3) *If* S *is the Klein bottle, then* $\mathcal{D}_o(S)$ *has* SO(2) *as strong deformation retract.*

4) *In all other cases* $\mathcal{D}_o(S)$ *is contractible.*

The case of the sphere was first established by Smale [57], using different methods. Corresponding results for compact surfaces with boundary have also been obtained [29].

If S *has Euler characteristic* $e(S) < 0$, *then* $\mathcal{D}_o(S)$

has no elements of finite order. In particular, $I \in \mathcal{D}_o(S)$ is the only compact Lie subgroup. On the other hand, $\mathcal{D}_o(S)$ contains arbitrarily large finite dimensional connected abelian subgroups.

(B) Standard topological properties of fibre bundles [41,58] imply that *in Case 4 above, all fibre bundles over a paracompact base space with structural group $\mathcal{D}_o(S)$ are topologically trivial.*

There are many interesting fibre spaces having Teichmüller spaces as base. On one hand, since $\mathfrak{J}(g)$ is a contractible Stein manifold, we are in a position to apply Grauert's theorem [36]: *Every holomorphic fibre bundle (with finite dimensional complex Lie structural group, connected or not) over $\mathfrak{J}(g)$ is holomorphically trivial.* In particular, the tangent vector bundle $T\,\mathfrak{J}(g)$ is such a bundle with group $GL(\mathbb{C}^{3g-3})$; therefore $\mathfrak{J}(g)$ is complex parallelizable.

On the other hand, there are complex fibre spaces, not having complex Lie structural groups, which arise in practice. For instance,

1) The universal bundle Φ: $\mathcal{M}(g) \to \mathfrak{J}(g)$ does not have a holomorphic section if $g \geq 2$ [24], although it does if $g = 1$ [28].

2) Let $S^{(n)}$ denote the n-fold cartesian product of S with itself. Then $\mathcal{D}_o(g)$ acts on $S^{(n)}$, and we can form the associated $(\mathcal{D}_o(g), S^{(n)})$ - bundle

$$W^{(n)}(g) = \mathcal{M}(g) \times_{\mathcal{D}_o(g)} S^{(n)} \to \mathfrak{J}(g);$$

see [13] for discussion. In particular, taking $n = 1$ gives the family of Riemann surfaces of genus g over $\mathfrak{T}(g)$.

3) Let $S_{(n)}$ denote the symmetric product of n copies of S; then $S_{(n)}$ is identified with the manifold of positive divisors of degree n on S. Again, $\mathcal{D}_0(S)$ induces an action on $S_{(n)}$, and we can form the complex fibre space

$$W_{(n)}(g) = \mathfrak{M}(g) \times_{\mathcal{D}_0(S)} S_{(n)} \to \mathfrak{T}(g).$$

See [32,54] for applications.

(C) The theorem of Kravetz that every finite subgroup of the modular group $\Gamma(g)$ has a fixed point has a striking topological interpretation [43]. Recall that the action of $\Gamma(g) = \mathcal{D}(g)/\mathcal{D}_0(g)$ on $\mathfrak{T}(g)$ is the quotient by $\mathcal{D}_0(g)$ of the action of $\mathcal{D}(g)$ on $\mathfrak{M}(g)$. If $\Phi\colon \mathfrak{M}(g) \to \mathfrak{T}(g)$ and $\theta\colon \mathcal{D}(g) \to \Gamma(g)$ are the quotient maps, then the subgroup of $\Gamma(g)$ which fixes $\Phi(J)$ is the *isomorphic* image under θ of the subgroup of $\mathcal{D}(g)$ fixing J. Thus Kravetz's theorem implies that *every finite subgroup of $\Gamma(g)$ is the isomorphic image of a subgroup of $\mathcal{D}(g)$*. Applying this to a cyclic group, we obtain a theorem of Nielsen [49]; see also MacBeath [46]: *Let $f \in \mathcal{D}(g)$. If the iterate f^n is homotopic to the identity, there exists f_0 homotopic to f with f_0^n = identity.*

(D) The following is an application of Teichmüller theory to differential geometry.

Theorem. *Any closed Riemann surface S is conformally equivalent to a smoothly embedded surface in $\mathbb{R}^3$. In fact, we can suppose that the Euclidean model is the locus of zeros of a real polynomial.*

The first assertion was proved by Garsia-Rodemich [35] in case genus(S) = 1, using methods of quasiconformal mapping. The general case was established by Garsia [33,34]; he used the holomorphic structure of $\mathfrak{T}(g)$, and Teichmüller's extremal mapping.

References

[1] L. V. Ahlfors, The complex analytic structure of the space of closed Riemann surfaces. Analytic Functions. Princeton 1960, 45-66.

[2] ————, Some remarks on Teichmüller's space of Riemann surfaces. Ann. of Math. 74 (1961), 171-191.

[3] ————, Curvature properties of Teichmüller's space. J. Analyse Math. 9 (1961), 161-176.

[4] ————, Teichmüller's spaces. Proc. Inter. Cong. Math. Stockholm 1962, 3-9.

[5] ————, Lectures on quasiconformal mappings. Mathematical Studies No. 10. Van Nostrand. 1966.

[6] L. V. Ahlfors and L. Bers, Riemann's mapping theorem for variable metrics. Ann. of Math. 72 (1960), 385-404.

[7] S. I. Al'ber, On n-dimensional problems in the calculus of variations in the large. Soviet Math. 5^1 (1964), 700-704.

[8] A. Andreotti, On a theorem of Torelli. Amer. J. Math. 80 (1958), 801-828.

[9] M. A. Armstrong, The fundamental group of the orbit space of a discontinuous group. Proc. Camb. Phil. Soc. 64 (1968), 299-301.

[10] R. Baer, Kurventypen auf Flächen. J. reine angew. Math. 156 (1927), 231-246.

[11] ————, Isotopie von Kurven auf orientierbaren,

[12] W. L. Baily, Jr., On the moduli of Jacobian varieties. Ann. of Math. 71 (1960), 303-314.

[13] L. Bers, Spaces of Riemann surfaces. Proc. Inter. Cong. Math. 1958, 349-361.

[14] ——————, Quasiconformal mappings and Teichmüller's theorem. Analytic Functions. Princeton 1960, 349-361.

[15] ——————, Correction to "Spaces of Riemann surfaces as bounded domains", Bull. Am. Math. 67 (1961), 465-466.

[16] ——————, Simultaneous uniformization. Bull. Am. Math. Soc. 66 (1960), 94-97.

[17] ——————, Inequalities for finitely generated Kleinian groups. J. Analyse Math. 18 (1967), 23-41.

[18] ——————, On moduli of Riemann surfaces. Notes at E.T.H. Zürich 1964.

[19] L. Bers and L. Ehrenpreis, Holomorphic convexity of Teichmüller spaces. Bull. Am. Math. Soc. 70 (1964), 761-764.

[20] H. Busemann, Spaces with non-positive curvature. Acta Math. 80 (1948-49), 259-310.

[21] H. Cartan, Quotient d'un espace analytique par un groupe d'automorphismes. Algebraic geometry and topology. Princeton 1957, 90-102.

[22] C. J. Earle, The Teichmüller spaces for an arbitrary Fuchsian group. Bull. Am. Math. Soc. 70 (1964), 699-701.

[23] ——————, The contractibility of certain Teichmüller spaces.

[24] ————, On holomorphic cross-sections in Teichmüller spaces.

[25] C. J. Earle and J. Eells, On the differential geometry of Teichmüller spaces. J. Analyse Math. 19 (1967), 35-52.

[26] ————, The diffeomorphism group of a compact Riemann surface. Bull. Am. Math. Soc. 73 (1967), 557-559.

[27] ————, Foliations and fibrations. J. Diff. Geometry 1 (1967), 33-41.

[28] ————, A fibre bundle description of Teichmüller theory

[29] C. J. Earle and A. Schatz, Teichmüller theory for surfaces with boundary.

[30] J. Eells and J. H. Sampson, Harmonic mappings of Riemannian manifolds. Am. J. Math. 86 (1964), 109-160.

[31] D. B. A. Epstein, Curves on 2-manifolds and isotopies. Acta Math. 115 (1966), 83-107.

[32] H. Farkas, Special divisors and analytic subloci of Teichmüller space. Am. J. Math. 88 (1966), 881-901. Also, 89 (1967), 225.

[33] A. M. Garsia, An imbedding of closed Riemann surfaces in Euclidean space. Comm. Math. Helv. 35 (1961), 93-110.

[34] ————, On the conformal type of algebraic surfaces of Euclidean space. Comm. Math. Helv. 37 (1962-63), 49-60.

[35] A. M. Garsia and E. Rodemich, An embedding of Riemann surfaces of genus one. Pac. J. Math. 11 (1961), 193-204.

[36] H. Grauert, <u>Analytische Faserungen über holomorph-vollstandigen Räumen</u>. Math. Ann. 135 (1958), 263-273.

[37] A. Grothendieck, <u>Techniques de construction en géométrie analytique</u>. Sém. H. Cartan. E. N. S. (1960/1), Exposé 7.

[38] P. Hartman, <u>On homotopic harmonic maps</u>. Canad. J. Math. 19 (1967), 673-687.

[39] E. Heinz, <u>Über gewisse elliptische Systeme von Differentialgleichungen</u>. Math. Ann. 131 (1956), 411-428.

[40] A. Hurwitz, <u>Über algebraische Gebilde mit eindeutigen Transformationen in sich</u>. Math. Ann. 41 (1893), 403-442.

[41] D. Husemoller, Fibre Bundles. McGraw Hill 1966.

[42] K. Kodaira and D. C. Spencer, <u>On deformations of complex analytic structures I, II</u>, Ann. of Math. 67 (1958), 328-466.

[43] S. Kravetz, <u>On the geometry of Teichmüller spaces and the structure of their modular groups</u>. Ann. Acad. Sci. Fenn. 278 (1959), 1-35.

[44] H. Lewy, <u>On the non-vanishing of the Jacobian in certain one-to-one mappings</u>. Bull. Am. Math. Soc. 42 (1936), 689-692.

[45] W. B. R. Lickorish, <u>A finite set of generators for the homeotopy group of a 2-manifold</u>. Proc. Camb. Phil. Soc. 60 (1964), 769-778. Also, 62 (1966), 679-681.

[46] A. M. MacBeath, <u>On a theorem of J. Nielsen</u>. Quart. J. Math. Oxford 13 (1962), 235-236.

[47] W. Magnus, Über Automorphismen von Fundamentalgruppen barandeter Flächen. Math. Ann. 109 (1933/4), 617-646.

[48] D. Mumford, Abelian quotients of the Teichmüller modular group. J. Analyse Math. 18 (1967), 227-244.

[49] J. Nielsen, Abbildungsklassen endlicher Ordnung. Acta Math. 75 (1943), 23-115.

[50] H. E. Rauch, On the transcendental moduli of algebraic Riemann surfaces. Proc. Nat. Acad. Sci. 41 (1955), 42-49.

[51] ————, On the moduli of Riemann surfaces. Proc. Nat. Acad. Sci. 41 (1955), 236-238.

[52] ————, Weierstrass points, branch points, and the moduli of Riemann surfaces. Comm. Pure Appl. Math. 12 (1959), 543-560.

[53] ————, Singularities of the modulus space. Bull. Am. Math. Soc. 68 (1962), 390-394.

[54] ————, A transcendental view of the space of algebraic Riemann surfaces. Bull. Am. Math. Soc. 71 (1965), 1-39.

[55] J. H. Sampson, (to be published).

[56] K. Shibata, On the existence of a harmonic mapping. Osaka Math. J. 15 (1963), 173-211.

[57] S. Smale, Diffeomorphisms of the 2-sphere. Proc. Am. Math. Soc. 10 (1959), 621-626.

[58] N. E. Steenrod, The topology of fibre bundles. Princeton 1951.

[59] O. Teichmüller, Extremale quasikonfome Abbildungen und quadratische Differentiale. Abh. Preuss. Akad. Wiss. Math.-Nat. Kl. 22 (1939).

[60] A. Weil, Modules des surfaces de Riemann. Sém. Bourbaki. 1958. Exp. 168.

[61] H. H. Wu, Normal families of holomorphic mappings. Acta Math. 119 (1967), 193-233.

Global Stability Questions in Dynamical Systems

by S. Smale

We begin by giving two very old examples of dynamical systems exhibiting certain kinds of stability. Let M be a Riemanian manifold, smooth and compact for simplicity. Suppose $f: M \to R$ is a smooth function. Then for each $x \in M$, the derivative $Df(x): T_x(M) \to R$ is a linear function on the tangent space, defining a smooth 1-form on M. The Riemannian metric converts this to a vector field on M, the gradient of f or grad f. Let $X = -$ grad f and φ_t, t real, be the 1-parameter group or flow generated by X. These dynamical systems, the gradient dynamical systems are among the best understood. They have the property that if x is not a critical point of f, then $f(\varphi_t(x))$ decreases as t increases.

In general, a fixed point $p \in M$ of a dynamical system $\varphi_t: M \to M$ is said to be asymptotically stable or a sink, or attractor if there is a neighborhood $U = U(p)$ and that for $x \in U$, $\varphi_t(x) \to p$ as $t \to \infty$. Then it is an old and easily checked fact that if p is an isolated minimum of f, it is asymptotically stable for the gradient dynamical system defined by $-$ grad f.

Our second example is a closed orbit. Asymptotic stability of a closed orbit of a dynamical system is defined just as for a fixed point. To give a criterion for stability of a closed orbit β of a dynamical system (X, φ_t) on M take a submanifold Σ of codimension 1 meeting a point p of β transversally (submanifolds W, V of M <u>meet transversally</u> if at each point $x \in W \cap V$, the tangent spaces $T_x(W)$, $T_x(V)$ span $T_x(M)$). Define a diffeomorphism in a neighborhood V of p in Σ, $f: V \to \Sigma$, $f(p) = p$, by letting $f(q)$ be the first point on the orbit of q in Σ with $t > 0$. The derivative of f at p is a linear automorphism $Df(p): T_p(\Sigma) \to T_p(\Sigma)$ with say eigenvalues λ_i. It has been well-known for a long time that if $|\lambda_i| < 1$ all i, then β is asymptotically stable. One may see the plausibility of this in the following way. If all the eigenvalues have absolute value less than 1, then $Df(p)$ is a contraction and in a neighborhood N of p, f itself contracts points of Σ towards p. Then by the definition of f, $\varphi_t(q) \to \beta$ as $t \to \infty$ for q in N. Thus β is an attractor.

The second example also possesses a second kind of stability that falls under the general heading of structural stability. This has to do with the perturbing of the dynamical system itself. If in fact X is perturbed by a small amount in the C^1 sense, to X^1 say, then there still exists a closed orbit β^1 relative to X^1 and β^1 is an attractor for X^1.

The first example will have also this second stability property near p provided we make the following nondegeneracy assumption. The derivative $D\varphi_1(p)$ of the dynamical system at time 1, as an automorphism of the linear space $T_p(M)$ has only eigenvalues with absolute value less than 1.

Systems possessing this second kind of stability are important for mathematics of engineering where one only knows the dynamical system up to a small perturbation, ie., the coefficients of the differential equation are only given approximately.

These two kinds of stability, asymptotic and structural, are related in a deep way in general; very roughly speaking, asymptotic holding in a systematic way implies structural stabilit We restrict ourselves from now on to discrete dynamical systems, i.e. a dynamical system given by powers of a diffeomorphism $f:M \to M$, f^m, $m \in Z$; we think of m as a discrete form of time. In this case, the technical problems are a little simpler but the basic ideas are the same and the results extend to differential equations. One relation between discrete dynamical systems and differential equations appears in the second example. Furthermore every example of a discrete dynamical system can be "suspended" to an example of a differential equation in one higher dimension with corresponding properties. (See "Differentiable Dynamical Systems", Bull. AMS, November 1967, for this construction as well as general background material and references to the literature; we leave details of some of the constructions, as well as all history to this survey). With the above in mind, we turn now to a third example, which is newer and a much more complicated attractor, and indicates the sort of thing that a general theory must deal with.

The example is given as a map $f\colon D^2 \times S^1 \to D^2 \times S^1$ of the product of a 2-disk and a circle, a solid torus, into itself, which is a diffeomorphism onto its image. We ask that f contract on D^2, expand on S^1, winding around itself twice so that the image of $f(D^2 \times S^1) \subset D^2 \times S^1$ looks like that of our sole figure.

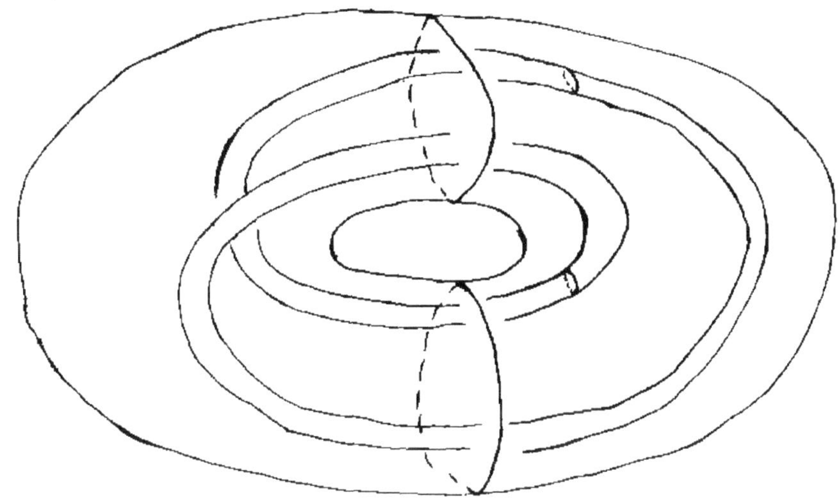

Then $f^2(D^2 \times S^1)$ winds around four times in $D^2 \times S^1$ and the reader will be able to see that $\alpha = \bigcap_{m>0} f^m(D^2 \times S^1)$ will be topologically a solenoid, locally a product of an interval and a Cantor set.

This "solenoid" α can be represented in the following way. Let $g\colon S^1 \to S^1$ be defined by $g(z) = z^2$ where S^1 is the set of complex numbers of absolute value 1. Then the inverse limit of these spaces relative to $g, g^2, g^3, \ldots$ is α and the inverse limit map is equivalent to f restricted to α. Considering α as a group, $f\colon \alpha \to \alpha$ is a group automorphism.

By the very construction, α is an attractor. In fact we are able to say, asymptotically at least, exactly what happens to any $x \in D^2 \times S^1$ under positive iteration. Given $x \in D^2 \times S^1$, there is $p \in \alpha$ such that $d(f^m(x), f^m(p)) \to 0$ as $m \to \infty$. Since f on α is given algebraically, this gives us a good description of f even in a neighborhood of α.

Furthermore, all these properties of this example remain true for any perturbation f^1 of f. Thus the example can't be called artifical; it is necessary to encompass this in any general study of differential equations.

We now make precise this second kind of stability, the structural stability of these examples.

For M a smooth compact manifold, $\mathrm{Diff}(M)$ will be the space of all diffeomorphisms of M (ie. generators of discrete dynamical systems) with the topology of C^1 uniform convergence. Diffeomorphisms f, g are called <u>topologically conjugate</u> if there exists a homeomorphism $h\colon M \to M$ such that $fh = hg$. In this case f and g have the same orbit structure and are the same from the point of view of dynamical systems. Then $f \in \mathrm{Diff}(M)$ is called structurally stable if there exists a neighborhood $N(f)$ in $\mathrm{Diff}(M)$ such that for $f^1 \in N(f)$, f^1 is topologically conjugate to f. Thus f is structurally stable if when perturbed, it has the same qualitative behavior.

A second weaker but important kind of stability, Ω-stability is defined as follows. First, the non-wandering set $\Omega = \Omega(f)$ of f in Diff(M) is defined as the set of $x \in M$ such that any neighborhood U of x has the property: there exists $m > 0$ with $f^m(U) \cap U \neq \emptyset$. Thus Ω is closed, invariant and consists of points of M with the mildest possible kind of recurrence. The non-wandering set is where the action is and thus what happens on Ω is the most important aspect of a dyanamical system.

Say that $f \in$ Diff(M) is <u>Ω-stable</u> if there is a neighborhood N(f) with the property that for $f^1 \in N(f)$, there exists a homeomorphism h: $\Omega(f) \to \Omega(f^1)$ such that for $x \in \Omega(f)$ $fh(x) = hf^1(x)$. Clearly structural stability implies Ω-stability.

We pose now the questions: Find necessary and sufficient conditions for a dynamical system $f \in$ Diff(M) to be

(A) Structurally stable

(B) Ω-stable

Of course for this to be very meaningful and important, these necessary and sufficient conditions should be "practical" (rather than tautological).

We have conjectured answers to these questions and a solid part of these conjectures have been proved true.

The main condition in both cases is what I have called Axiom A (for a diffeomorphism) and this condition goes as follows: For $f \in$ Diff(M), the derivative Df is a bundle automorphism of the tangent bundle T(M). Let $T_\Omega(M)$ denote the part of this tangent bundle over Ω.

Then Axiom A is satisfied for $f \in \text{Diff}(M)$ if:

Axiom A: The periodic points are dense in $\Omega(f)$ and Df restricted to $T_\Omega(M)$ splits into a sum of a contracting and an expanding automorphism of invariant subbundles E^s, E^u respectively.

In this case $T_\Omega(M) = E^s + E^u$, E^s is mapped onto E^s, E^u onto E^u by Df. That $E^s \to E^s$ is contracting means that for some (and hence any) Riemannian metric, the following estimate is satisfied:

all $x \in \Omega$, $v \in E^s$, $\|Df^m(x)(v)\| \leq C\lambda^m\|v\|$, $C > 0$, $0 < \lambda < 1$.

Finally a bundle automorphism is expanding if its inverse is contracting.

If each $x \in \Omega$ is fixed then Axiom A amounts to saying, for each x, $Df(x): T_x(m) \to T_x(M)$ has all eigenvalues not 1 in absolute value. The eigenspaces give the desired splitting. Also α in the 3rd example has this "hyperbolic" structure, so Axiom A is satisfied here too.

If f satisfies Axiom A and $p \in \Omega$, denote by $W^s(p)$ the set of $x \in M$ such that the distance $d(f^m(x), f^m(p)) \to 0$ as $m \to \infty$. Then $W^s(p)$ is a 1-1 immersed cell in M called the stable manifold of p. The unstable manifold $W^u(p)$ is defined as the stable manifold of p for f^{-1}.

Then our second condition is:

Strong Transversality Condition For each $x, y \in \Omega$, $W^s(x)$, $W^u(y)$ meet transversally.

Conjecture A: A discrete dynamical system $f \in \text{Diff}(M)$ is structurally stable if and only if f satisfies Axiom A and the Strong Transversality Condition.

This conjecture has been proved recently in the special case that Ω is finite.

<u>Theorem</u>. (Palis-Smale) If $f \in \mathrm{Diff}(M)$ and $\Omega(f)$ is finite then f is structurally stable if and only if f satisfies Axiom A and the Strong Transversality Condition.

A similar theorem is also valid for the differential equation case.

For a gradient dynamical system, φ_t generated by $X = -\operatorname{grad} f$, $\varphi_1 : M \to M$ has $\Omega(\varphi_1)$ equal to the set of critical points of f. Also, almost all f have isolated critical points and in fact, one gets as a corollary of the above theorem

<u>Corollary</u>. Almost all gradient dynamical systems are structurally stable.

<u>Corollary</u>. Every compact M admits a structurally stable dynamical system.

Now let us consider the situation for Ω-stability. For this we remark first that a theorem, "the spectral decomposition theorem" asserts that if $f \in \mathrm{Diff}(M)$ satisfies Axiom A, then there is a canonical finite decomposition into disjoint closed invariant sets, $\Omega(f) = \Omega_1 \cup \cdots \cup \Omega_k$ such that on each Ω_i, f has a dense orbit.

Define now $W^s(\Omega_i) = \bigcup_{x \in \Omega_i} W^s(x)$, $W^u(\Omega_i) = \bigcup_{x \in \Omega_i} W^u(x)$ and a

<u>cycle</u> to be a sequence $\Omega_{i_1}, \cdots, \Omega_{i_k}$, $k > 1$ of distinct Ω_{i_m} such that $W^u(\Omega_{i_m}) \cap W^s(\Omega_{i_{m+1}}) \neq \emptyset$ $m = 1, \cdots, k$, $\Omega_{i_{k+1}} = \Omega_{i_1}$. Then for f satisfying Axiom A, the no cycle property holds if:

<u>No cycle property</u>: For f, there are no cycles.

Then we can formulate our 2nd conjecture.

<u>Conjecture B</u>. If $f \in \mathrm{Diff}(M)$ is a discrete dynamical system, then f is Ω-stable if and only if it satisfies Axiom A and the no cycle property.

In one direction this has been proved.

<u>The Ω-stability Theorem:</u> If f satisfies Axiom A and the no cycle property then it is Ω-stable.

Also J. Palis pointed out to me that if f satisfies Axiom A and is Ω-stable, then it also satisfies the no cycle property. We have the picture

Structural Stability $\Longrightarrow$ Ω-stability

and assuming Axiom A

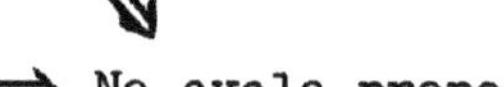

Strong Transversality Condition $\Longrightarrow$ No cycle propert

The important question coming out of this discussion is: Does Ω-stability imply Axiom A?

Titles and abstracts of the lectures presented in this series other than those contained in this volume are listed below. They will appear in subsequent volumes.

Modern Methods and New Results in Complex Analysis

Professor Paul Malliavin, University of Paris

Entire Functions of Exponential Type and Harmonic Analysis

Support of a distribution - Convolution equation - Closure of characters - Interpolation and extremal problems.

Banach Algebras and Applications

Professor Andrew M. Gleason, Harvard University

The Maximal Ideal Space of a Function Algebra

Fifteen years ago it was thought that the maximal ideal space of a function algebra must be in some sense analytic. Since then counter examples have been given by Stolzenberg, Kallin, Garnett, and others. Positive progress has been made, primarily in the one dimensional case, by Wermer, Hoffman, and others, but the relation between a function algebra and its maximal ideal space remains shrouded in mystery.

Geometric and Qualitative Aspects of Analysis

Professor I. M. Singer, Massachusetts Institute of Technology

The Laplacian on G-structures

The Laplacian on a G-structure can be expressed in terms of the covariant differential and an explicit function of the curvature of the G-structure. This unifies some results on invariant subspaces of the Laplacian, and vanishing theorems. It also allows for a more direct computation of the index in some special cases.

Analysis and Representation Theory

Professor Harish-Chandra, Institute for Advanced Study

Some Applications of the Schwartz Space of a Semisimple Lie Group

Let G be a semisimple Lie group. The analogue for G of the usual Schwartz space (on R^n) plays an important role in the harmonic analysis on G.

Professor Richard V. Kadison, University of Pennsylvania

Automorphisms of Operator Algebras

Recent theorems on groups of automorphisms of C* and von Neumann algebras will be discussed together with their relation to modern physical theory.

Professor George W. Mackey, Harvard University

Some Applications of Double Coset Intertwining Operators

Definition of induced representation and intertwining operator. Intertwining operators for induced representations of finite groups and their connection with double cosets. Intertwining operators defined by double cosets in general. Properties of such operators in certain special cases. Instances of the occurrence of double coset intertwining operators in integral geometry, in the theory of harmonic functions and in the theory of automorphic forms.

Modern Analysis and New Physical Theories

Professor Irving E. Segal, Massachusetts Institute of Technology

Mathematical Theory of Quantum Fields

A contemporary mathematical view of quantum field theory, with answers for such questions as: what is a quantum field?, and illustrations from quantum electrodynamics. The formulation of the S-matrix as the "time-ordered exponential of the interaction Hamiltonian" will be discussed.

Professor James Glimm, Massachusetts Institute of Technology and New York University

Foundations of Quantum Field Theory

Mathematical problems presented by quantum field theory. The occurrence of infinities and the features of the physics responsible for their presence. Approximations to the correct laws of physics which remove the infinities. Removal of all approximations to obtain as a limit a quantum field Φ with a Φ^4 selfinteraction in two dimensional space time.

Professor David Ruelle, Institut des Hautes Études Scientifique

New Methods and Problems in Statistical Mechanics

Recent developments in the statistical mechanics of infinite systems will be discussed. Some insight into the meaning of the Gibbs phase rule is obtained.

Modern Harmonic Analysis and Applications

Professor Elias M. Stein, Princeton University

Variations on the Littlewood-Paley Theme

The purpose of this lecture will be to discuss a new approach to the classical Littlewood-Paley theory, viewed in the framework of symmetric diffusion semi-groups. This leads to a generalization of parts of that theory and further applications in a variety of new situations.

Professor R. P. Langlands, Yale University

Problems in the Theory of Automorphic Forms

Analogues, for an automorphic form on a general reductive group, of the Hecke L-series will be introduced. Some questions in the theory of group representations to which they give rise will be discussed.

Professor Bertram Kostant, Massachusetts Institute of Technology

On Orbits and Representations of Lie Groups

We are concerned with the existence and irreducibility of unitary representations of Lie groups. Our general approach to unitary representations is by means of a quantization theory associated with certain symplectic manifolds. L. Auslander and I successfully applied this theory to obtain criteria for sovable Lie groups to be of type I and to determine all the irreducible unitary representations of such groups.

Integration in Function Spaces and Applications

Professor Leonard Gross, Cornell University

Potential Theory on Hilbert Space

Basic definitions and concepts in the theory of integration over infinite dimensional vector spaces will be described. Newtonian potential theory on Euclidean n-space will be extended to potential theory on an infinite dimensional Hilbert space as an application and motivation. It will be shown how Wiener space arises from this considerations.

Professor Richard M. Dudley, Massachusetts Institute of Technology

Random Linear Functionals

Given a linear mapping of a topological vector space S into random variables, when does it define a countably additive probability measure on the dual space S'? There are recent results of Laurent Schwartz and co-workers. The Gaussian case leads to questions about measuring the sizes compact, convex subsets of Hilbert space.

Professor Jacob Feldman, University of California, Berkeley

Some Aspects of Gaussian and Decomposable Processes

Discussion of conditions for absolute continuity of measures induced by various classes of stochastic processes. Transformations of such processes. The notion of continuous product measure,and the study and classification of decomposable processes from this viewpoint.